THE GRAPHIC GUIDE TO BEEKEEPING

Yves Gustin

THE GRAPHIC GUIDE TO BEEKEEPING

Your Complete Visual Resource for Sweet Success

4880 Lower Valley Road • Atglen, PA 19310

Other Schiffer Books on Related Subjects:

Leatherwork: Traditional Handcrafted Leatherwork Skills and Projects, Nigel Armitage, ISBN 978-0-7643-6039-8

The Global Outdoor Survival Guide: Basic to Advanced Skills for Every Environment, Joe Vogel, ISBN 978-0-7643-5426-7

Building Dry-Stack Stone Walls, Rob Gallagher, Sean Malone & Joe Piazza, ISBN 978-0-7643-3056-8

© 2021 by Schiffer Publishing Ltd., English edition

Translated from the French by Simulingua, Inc. Originally published as *L'apiculture en bande dessinée*, © 2008, 2017, Éditions Rustica, Paris

Library of Congress Control Number: 2020943417

All rights reserved. No part of this work may be reproduced or used in any form or by any means—graphic, electronic, or mechanical, including photocopying or information storage and retrieval systems—without written permission from the publisher.

The scanning, uploading, and distribution of this book or any part thereof via the Internet or any other means without the permission of the publisher is illegal and punishable by law. Please purchase only authorized editions and do not participate in or encourage the electronic piracy of copyrighted materials.

"Schiffer," "Schiffer Publishing, Ltd.," and the pen and inkwell logo are registered trademarks of Schiffer Publishing, Ltd.

Cover design by Jack Chappell
Type set in DK Bocadillo/Cheddar Jack/CCComicrazy-Roman
Technical review assistance: Deborah Klughers, EAS Master Beekeeper, proprietor of Bonac Bees, www.BonacBees.com

ISBN: 978-0-7643-6124-1
Printed in China

Published by Schiffer Publishing, Ltd.
4880 Lower Valley Road
Atglen, PA 19310
Phone: (610) 593-1777; Fax: (610) 593-2002
E-mail: Info@schifferbooks.com
Web: www.schifferbooks.com

For our complete selection of fine books on this and related subjects, please visit our website at www.schifferbooks.com. You may also write for a free catalog.

Schiffer Publishing's titles are available at special discounts for bulk purchases for sales promotions or premiums. Special editions, including personalized covers, corporate imprints, and excerpts, can be created in large quantities for special needs. For more information, contact the publisher.

We are always looking for people to write books on new and related subjects. If you have an idea for a book, please contact us at proposals@schifferbooks.com.

INTRODUCTION

Since time immemorial, beekeeping has never changed—except for the tools. The methods I explain here stay relevant because they're the simplest. Everything you'll read in this book has been tested firsthand by me. After years of working with experienced beekeepers, and after reading everything I could get my hands on about the subject, I wanted to describe beekeeping as naturally and simply as possible.

Everything is explained in this book: Use this guide to build your own hives and manage them, starting by choosing a space that's sunny, protected, and with, of course, a variety of honey plants not too far from your hives. It's up to you to study the details and apply them as you observe and live your own discovery of the joys of bees.

Good luck and happy beekeeping!
Yves Gustin

CONTENTS

What Are Bees Used For?	10
The Life of a Bee	12
Anatomy of a Bee	16
The Life of a Drone	20
Anatomy of a Drone	22
The Life of a Queen	24
Anatomy of a Queen	28
The Sting and Venom	30
Bee Language	34
Color	38
Smell	41
Heat	42
The Hive	44
Hive Construction	48
The Activity of a Hive	52
Building a Frame	54
Installing Wire	56
Embedding Wax Foundation	58
Drinking Water	62
Cleaning the Apiary	64
Setting Up the Hives	68
Migratory Beekeeping	70
Predators	74
Some Nectar-Producing Plants to Grow	76
Some Sources of Varietal Honey Flavors	77
Beneficial Species for the Apiary	78
Grafting (for More Trees!)	80
Installing Honey Supers	82
Harvesting	84
Robbing	88
Storing Supers	90
Honey Processing	94
The Uncapping Tank	100
Fall Work	102
Making Bee Candy	04
Late Feeding	108
Preparing for Winter	110
Winter Hive Visit	112
A Clean Bottom Board	116
Winter Tasks	120
The Spring Visit	122
Making Syrup	124
The Swarming Period	126
Capturing a Swarm	130
Increasing Your Bees	134
The Artificial Swarm	138
Transferring a Hive	142
Breeding Queens	146
Introducing Queens	152
Joining	156
A Drone-Laying Colony	158
Honey	162
Selling Your Honey	168
Mead	172
Pollen	176
Royal Jelly	182
Wax	186
A Solar Wax Melter	190
Propolis	192
Poisoning	196
Diseases	200
Parasites	204
Wax Moths	210
Glossary	214

WHAT ARE BEES USED FOR?

WITHOUT BEES AND OTHER POLLINATING INSECTS, WE WOULD SEE ONLY THIS DESERT LANDSCAPE MADE OF SAND AND SOME OF THE TOUGHEST PLANTS. IN DESERTS, PLANTS DON'T FIND THE HUMUS THEY NEED TO SURVIVE. POLLINATING INSECTS POLLINATE PLANTS TO PRODUCE FLOWERS, FRUITS, AND LEAVES, WHICH EVENTUALLY FALL AND ROT ON THE GROUND, AND THAT BECOMES HUMUS.

YEAR AFTER YEAR, THIS ORGANIC AMALGAM BUILDS UP RAW AND GOOD-QUALITY HUMUS, WHERE BACTERIA, FUNGI, AND EARTHWORMS DEVELOP A FAVORABLE ENVIRONMENT FOR PLANT GROWTH.

WE CAN SAY THAT THE EARTH IS A LARGE DIGESTIVE TRACT (SIMILAR TO OURS) WHERE BILLIONS OF LIVING BEINGS ARE THE TOP CHEFS FOR PLANTS. THEY PREPARE SMALL MEALS FOR EACH PLANT AND SERVE NITROGEN FROM THE AIR AS A DESSERT.

THE PRESENCE OF BEES PROMOTES LIFE AND PROVIDES THE PLEASURE OF THE SEASONS, THANKS TO THEIR POLLINATION.

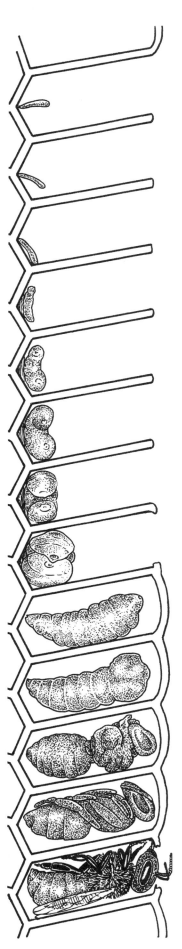

DAY 1: The translucent white egg, measuring 1.5 mm, is at the bottom of the cell, attached at its smallest end.

DAY 2: The egg begins to tilt.

DAY 3: The egg is lying at the bottom of the cell. During the first three days, the embryo develops inside the egg.

DAY 4: The larva comes out of the egg, and a nurse bee deposits a drop of royal jelly in which the larva will bathe and feed.

DAY 5: Constantly fed, the larva grows rapidly and begins to take on a curved shape.

DAY 6: The larva fills the bottom of the cell, and its two ends meet.

DAY 7: Nurses stop feeding the larva royal jelly, which is replaced by a porridge made of honey, pollen, and water.

DAY 8: The larva's diet consists of the same ingredients as before; however, the amount of pollen it's given increases more and more until capping.

DAY 9: The wax makers seal the cell containing the larva with a material made of wax, pollen, and debris; this cover allows air to pass through. The larva changes its position so that its head is directed toward the exit; from that moment on, it wraps itself in its silk cocoon, secreted by salivary glands.

DAY 10
DAY 11: Transformation of the larva into a pupa
DAY 12
DAY 13
DAY 14: Rest period

DAY 15: The pupa is complete.

DAY 16
DAY 17
DAY 18
DAY 19
DAY 20: Metamorphosis of the pupa into an insect

AS SPRING APPROACHES, THE QUEEN SETS TO WORK, AND IN EACH CELL YOU FIND AN EGG. LET'S TAKE A FERTILIZED EGG AS AN EXAMPLE AND SEE HOW IT EVOLVES.

THE METAMORPHOSIS

EGG

LARVA

LARVAL STAGE

PUPA

BEE

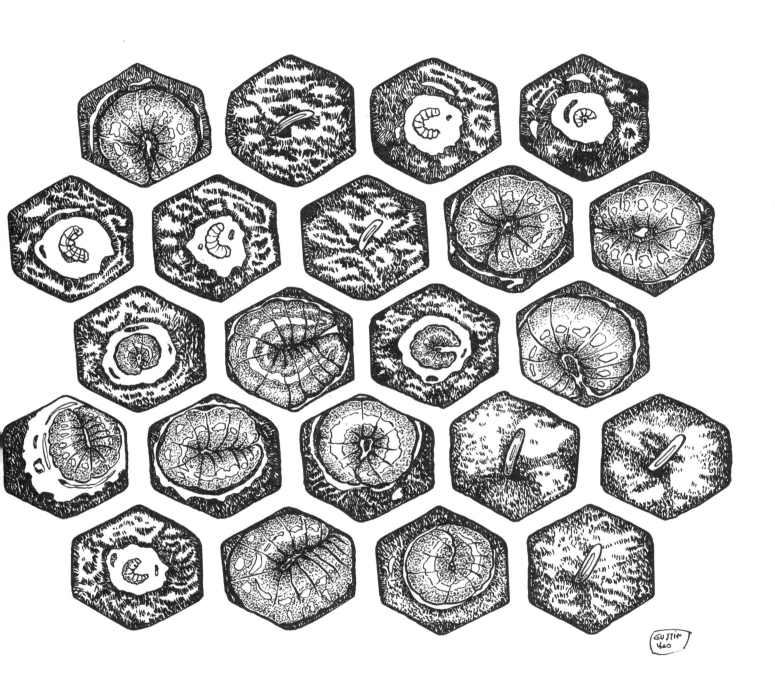

• THE LIFE OF A BEE •

ANATOMY OF A BEE

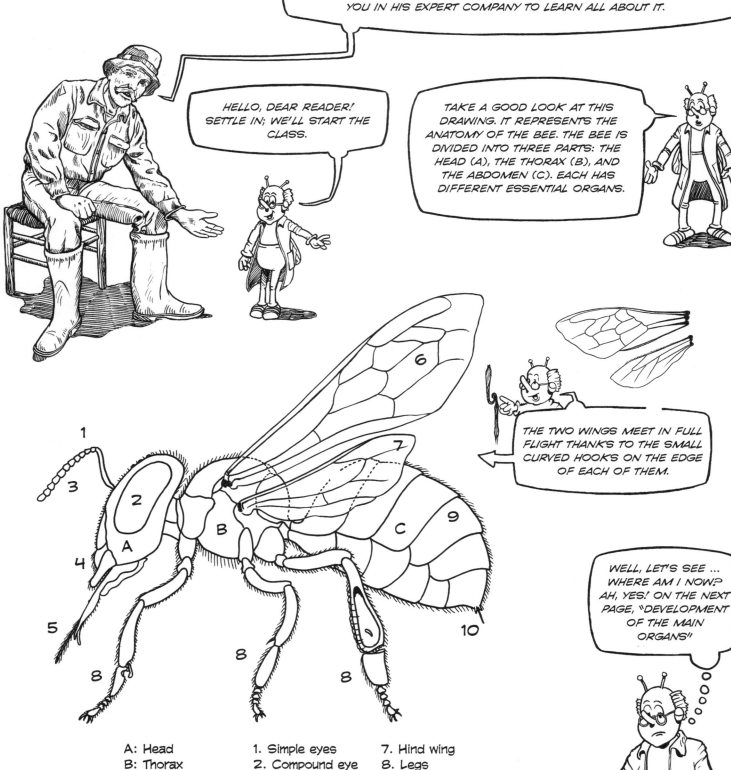

THE HEAD INCLUDES THE EYES, ANTENNAE, AND ORAL APPARATUS. LET US EXAMINE EACH OF THESE ELEMENTS IN DETAIL.

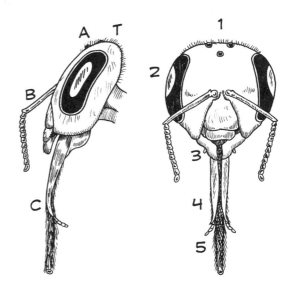

A: The eyes: there are two kinds.
1. Simple eyes: three small eyes in the center of the head and arranged in a triangle, allowing close vision in the dark and more particularly inside the hive.

2. Two large eyes with thousands of facets, used for distance and external vision (sensitive to ultraviolet rays).

B: Antennae: there are two of them. They are the sensory organs that allow bees to communicate with each other inside the hive.

C: The oral apparatus: it consists of three parts. 3. Two mandibles allow the bee to handle and chew wax, pollen, and propolis.
4. Two jaws serve to grind.

5. A tongue allows the bee to suck in the nectar; its length varies from 5 to 7 mm depending on the breed.

THE THORAX (B) CONSISTS OF THREE SECTIONS, EACH OF WHICH HAS A PAIR OF LEGS AND TWO PAIRS OF WINGS ATTACHED TO THE LAST TWO SECTIONS. A PARTICULAR FUNCTION IS ASSIGNED TO EACH PAIR OF LEGS.

THE FRONT LEGS (6) CLEAN THE ANTENNAE. THE MIDDLE LEGS (7) ARE EQUIPPED WITH A SPUR TO PASS POLLEN BALLS THROUGH THE BASKETS. THE REAR LEGS (8) ARE USED TO STORE POLLEN IN BASKETS (H) AND COLLECT THE WAX SCALES UNDER THE ABDOMEN.

THE ABDOMEN (C) IS MADE UP OF SEVEN SEGMENTS WHERE THEY ARE INTERTWINED: 9. THE LARGE WAX GLANDS, 10. THE NASONOV GLAND, 11. THE STINGER AND ITS VENOM GLANDS.

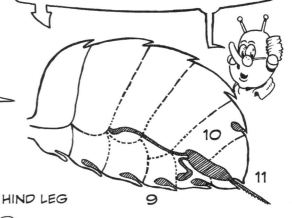

FORELEG

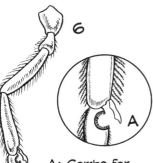

A: Combs for cleaning the antennae, the tongue, and the eyes

MIDDLE LEG

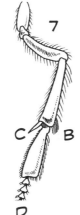

B: Comb
C: Spur
D: A type of suction cup (*Pulvillus*) that is under its hooks and allows the bee to adhere to smooth surfaces.

HIND LEG

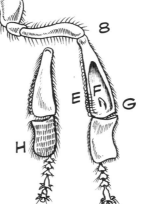

Interior   Exterior

E: Claws
F: Hook that holds the pollen in place
G: Pollen brush
H: Pollen basket

• ANATOMY OF A BEE •

NOW THAT WE HAVE SEEN THE OUTSIDE OF THE BEE, LET'S LOOK INSIDE AND GET TO THE HEART OF THE MATTER.

THE CIRCULATORY SYSTEM
The bee's blood is colorless because it does not contain red blood cells. The heart (A) sucks in and pushes the blood back to the head through the aorta (B); the blood is distributed throughout the body and returns to the abdomen, where it will be sucked in again.

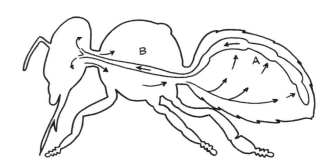

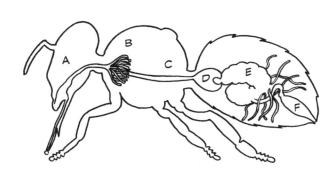

THE DIGESTIVE SYSTEM
It passes through the entire body of the bee. It is divided into seven parts.
    A: Pharynx (gullet)
    B: Salivary glands
    C: Esophagus
    D: Crop (honey stomach)
    Not pictured: Proventriculus (part of the stomach that stores nectar without digesting it)
    E: Midgut
    F: Large intestine (rectum)

THE RESPIRATORY SYSTEM
The bee doesn't breath through its mouth but through spiracles that allow air to penetrate into its trachea's air sacs.

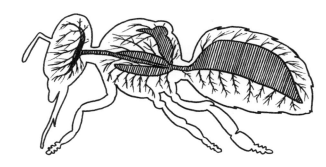

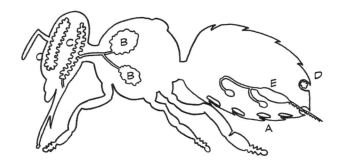

THE GLANDULAR SYSTEM
    A: Wax glands under the last four rings of the abdomen
    B: Salivary glands
    C: Hypopharyngeal gland
    D: Nasonov gland
    E: Venom glands

THE NERVOUS SYSTEM
The brain and ganglion chains control the bee.
  A: SIMPLE EYES
  B: ANTENNAE
  C: BRAIN
  D: THORACIC NODES
  E: ABDOMINAL NODE

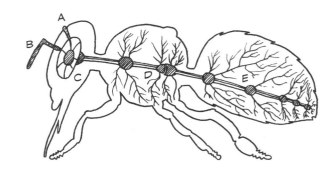

THE MUSCULAR SYSTEM
It is very developed, especially the wing muscles.
  A: DORSAL NERVURE

CHEST SECTION THAT SHOWS THE WING MUSCLES:
  B: LONGITUDINAL MUSCLES
  C: VERTICAL MUSCLES

• ANATOMY OF A BEE •

When the sunny days come, the queen works to fill the cells. According to her wishes, she lays fertilized eggs, which will become female bees, or unfertilized eggs, which will give birth to males, like me! Let's see how this egg evolves.

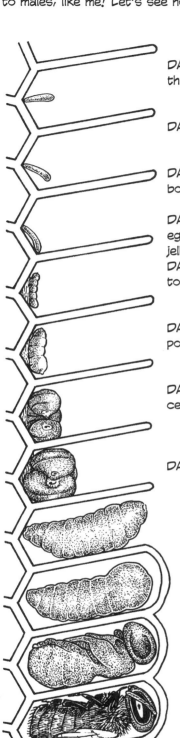

DAY 1: The egg is in the bottom of the cell.

DAY 2: The egg begins to tilt.

DAY 3: The egg is lying in the bottom of the cell.

DAY 4: The larva emerges from the egg; its food is made from royal jelly.

DAYS 5 & 6: The larva continues to grow.

DAY 7: Its diet consists of honey, pollen, and water.

DAY 8: The larva fills the entire cell.

DAY 9: The larva straighten up.

CAPPING

DAYS 10 TO 23: Transformation of the larva into a pupa.
After a period of rest, the metamorphosis continues and ends with a perfect insect.

DAY 24: The insect leaves the cell and hurries off to be fed by the workers.

FROM THE 1ST TO THE 12TH DAY OF ITS LIFE, THE DRONE HAS NO ESSENTIAL ACTIVITIES.

FROM THE 13TH DAY ON, HE'S ABLE TO IMPREGNATE A QUEEN, IF HE'S PART OF THE CONQUERING TEAM (4 OR 5 SUBJECTS OUT OF 2,000). AFTER THE PERFORMANCE OF HIS ACT IN THE SKIES, HE'S DOOMED. HE IS STRIPPED OF HIS REPRODUCTIVE ORGANS AND DIES.

AND DIE! WHAT DOES HE MEAN BY THAT?

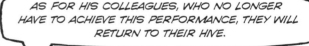

AS FOR HIS COLLEAGUES, WHO NO LONGER HAVE TO ACHIEVE THIS PERFORMANCE, THEY WILL RETURN TO THEIR HIVE.

THE LIFE SPAN OF A DRONE IS ABOUT THREE MONTHS, BUT THIS TIME VARIES ACCORDING TO THE SEASON AND POLLEN INPUTS.

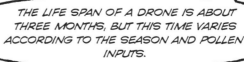

• THE LIFE OF A DRONE •

ANATOMY OF A DRONE

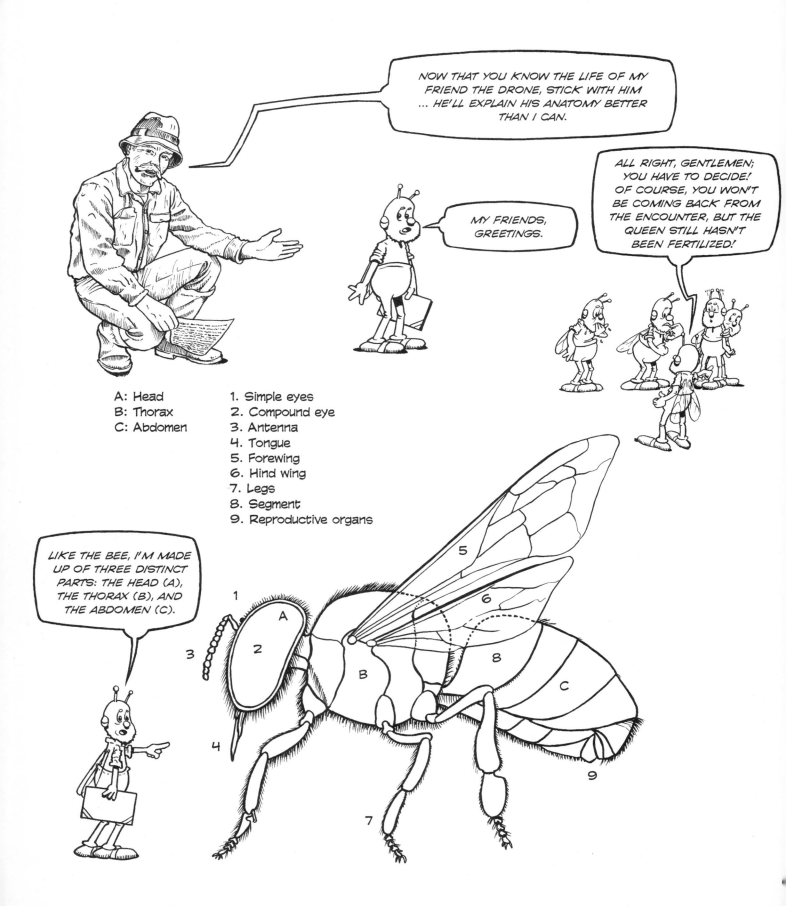

The head (A) includes:
1. Two very large eyes, placed on either side of the head. They contain a number of lenticular facets significantly superior to those of the queen and the workers.

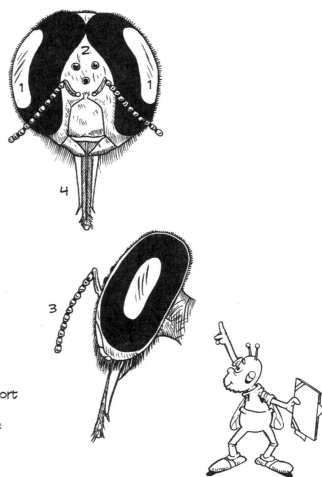

2. The simple eyes. Three small eyes in the center of the head are used for near vision.
3. The antennae that allow communication
4. The oral apparatus is deficient: the tongue, which is too short to prevent the drone from foraging and even feeding itself on honey inside the hive. The drone is therefore dependent on the workers.

The thorax (B) has two pairs of wings (5) and (6) and three pairs of legs (7), the latter without harvesting instruments.

The abdomen (C), like that of the queen or worker, is made up of segments (8). The drone has no stinger, but as compensation, it has reproductive organs (9), which give it the right to stay in the hive. Remember that without the drones, the queens wouldn't be fertilized and the colonies would die out.

1. Testis
2. Seminal vesicle
3. Mucus glands
4. Cornelius
5. Lobe
6. Bulb

THE LIFE OF A QUEEN

NOW THAT WE'VE STUDIED THE LIFE OF THE BEE AND THE DRONE, LET'S TAKE A CLOSER LOOK AT THE LIFE OF HER MAJESTY THE QUEEN (OR MOTHER).

ALTHOUGH I WAS DESTINED TO BECOME ONLY A WORKER, I WAS CHOSEN TO BE THE NEW QUEEN. NOW THAT I'M QUEEN, I CAN BE IDENTIFIED BY MY SIZE (0.71 TO 0.79 INCHES, OR 18 TO 20 MM), WHICH IS LARGER THAN THAT OF THE WORKERS (0.55 TO 0.59 INCHES, OR 14 TO 15 MM), AS WELL AS MY LEGS, ALSO LONGER. I CAN'T FORAGE OR PRODUCE WAX; THAT'S NOT MY JOB. I HAVE NO BASKETS ON MY LEGS, PLUS MY MANDIBLES AND TONGUE ARE TOO SHORT. ON THE OTHER HAND, NATURE HAS DECIDED ON MY MAIN FUNCTION: EGG LAYING. THAT'S WHY MY ABDOMEN AND GENITALS ARE VERY DEVELOPED. I'M ARMED WITH A STING THAT IS USED ONLY TO KILL A RIVAL AT THE TIME OF BIRTH (WHEN SEVERAL QUEENS ARE BORN).

THIS NEW QUEEN WAS BORN BECAUSE OF ONE OF THREE CAUSES: DEATH OF THE OLD QUEEN, SWARMING (NATURAL OR ARTIFICIAL), OR REMOVAL OF THE OLD QUEEN BY THE BEEKEEPER.

ONCE THE QUEEN IS FERTILIZED, SHE LAYS ABOUT 2,000 EGGS PER DAY IN PEAK EGG-LAYING SEASON. THE ENERGY SHE NEEDS TO ENSURE THIS LAYING IS PROVIDED BY HER COURT, WHICH "FORCE-FEEDS" HER WITH ROYAL JELLY. THE QUEEN MAY STOP LAYING EGGS FOR VARIOUS REASONS: COOLING OF THE HIVE, A SHOCK THAT FRIGHTENS THE QUEEN, ABSENCE OF NECTAR, ETC.

THE QUEEN, OR MOTHER, DOES NOT LAY EGGS IN ANY OLD CELL; SHE FIRST MAKES SURE THAT IT'S CLEAN AND BEGINS TO LAY EGGS IN THE CENTER OF THE COMB. PROGRESSIVELY, SHE WILL WIDEN THE CIRCLE; WHEN THE FRAME IS FULL, SHE'LL START AGAIN IN A NEW ONE.

HERE'S A BEAUTIFUL BROOD FRAME THAT IS IN THE CENTER OF THE HIVE. LET'S OBSERVE IT ... AROUND THE BROOD OF ALL AGES IS HONEY, AS WELL AS POLLEN ESSENTIAL FOR ITS PROTEIN SUPPLY.

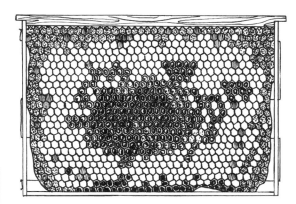

BROOD DISTRIBUTION
Frames #1 are intended to receive honey; frames #2 will also be filled with honey and pollen; frames #3 contain honey, pollen, and brood; frames #4 receive honey, pollen, and a lot of brood; frames #5, which are in the center of the hive, include a maximum of brood.

MY FRIEND THE BEEKEEPER SOMETIMES HAS SOME PROBLEMS FINDING ME, SO HE FOUND A WAY TO FIX IT BY COLOR-CODING ME. IT WORKS IN DIFFERENT WAYS: A COLORED DOT ON MY CHEST, A NUMBERED ADHESIVE TAG, ETC. IT ALSO ALLOWS HIM TO KNOW MY AGE. FIVE INTERNATIONAL COLORS HAVE BEEN CHOSEN FOR THIS PURPOSE. A QUEEN THAT WAS BORN IN 2020 WILL BE MARKED WITH A BLUE DOT, IN 2021 WITH A WHITE DOT, IN 2022 WITH A YELLOW DOT, IN 2023 WITH A RED DOT, IN 2024 WITH A GREEN DOT. THEN THIS CYCLE STARTS AGAIN WITH THE BLUE, AND SO ON.

IN MY LIFE, I CAN BE FERTILIZED SEVERAL TIMES.

MAYBE NEXT TIME I'LL HAVE THE HONOR ...

SOME BEEKEEPERS CUT OFF MY WINGS. THIS OPERATION IS CALLED CLIPPING AND IS INTENDED TO PREVENT ME FROM FLYING AWAY AT THE TIME OF SWARMING.

Queen marked with a colored dot

Cutting the wing along the line

I'M PART OF THE ROYAL COURT THAT WILL GUARD THE QUEEN FOR AS LONG AS SHE RELEASES THE PHEROMONE NECESSARY FOR THE PROPER FUNCTIONING OF THE COLONY. THIS QUEEN SUBSTANCE DISTRIBUTED AMONG THE BEES IS LEFT ON THE CELLS AFTER THE QUEEN'S PASSAGE.

IF THIS SUBSTANCE DECREASES OR BECOMES NONEXISTENT, THEN WE BUILD QUEEN CELLS TO REPLACE THE DEFICIENT OR MISSING QUEEN

THE QUEEN MUST BE REPLACED. LET'S OBSERVE THE DEVELOPMENT OF THE FUTURE QUEEN FROM A FERTILIZED EGG.

DAY 1: As with bees, the egg is at the bottom of the cell.
DAY 2: Attached by its smallest end, the egg begins its inclination of about 45°.
DAY 3: The egg is lying at the bottom of the cell.
DAY 4: The larva emerges from the egg, the nurses quickly feed it royal jelly, and the wax makers begin to build the queen cell.

 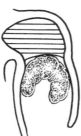  

DAY 5: The larva grows bigger, pampered by the nurses.
DAY 6: The larva fills the bottom of the cell.
DAY 7: While the nurses have stopped feeding royal jelly to the workers' larvae, our future queen will be fed exclusively royal jelly.
DAY 8: The construction of the queen cell is complete; the larva fills the entire cell.
DAY 9: Capping

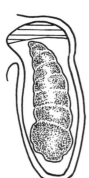 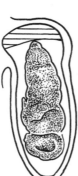   

A LOVELY QUEEN CELL

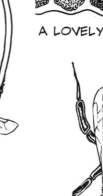

QUEEN

DAY 10: The larva surrounds itself with a silk cocoon.
DAYS 11 & 12: Rest period
DAY 13: Transformation into a pupa
DAYS 14 & 15: Metamorphosis of the pupa into a perfect insect
DAY 16: The queen's exit

26 • THE LIFE OF A QUEEN •

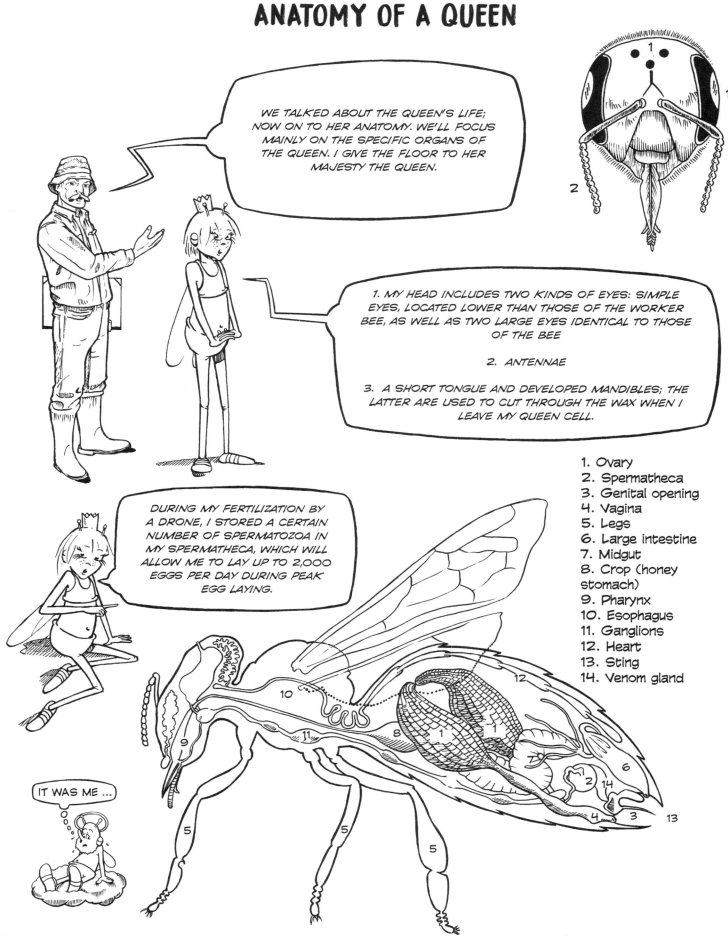

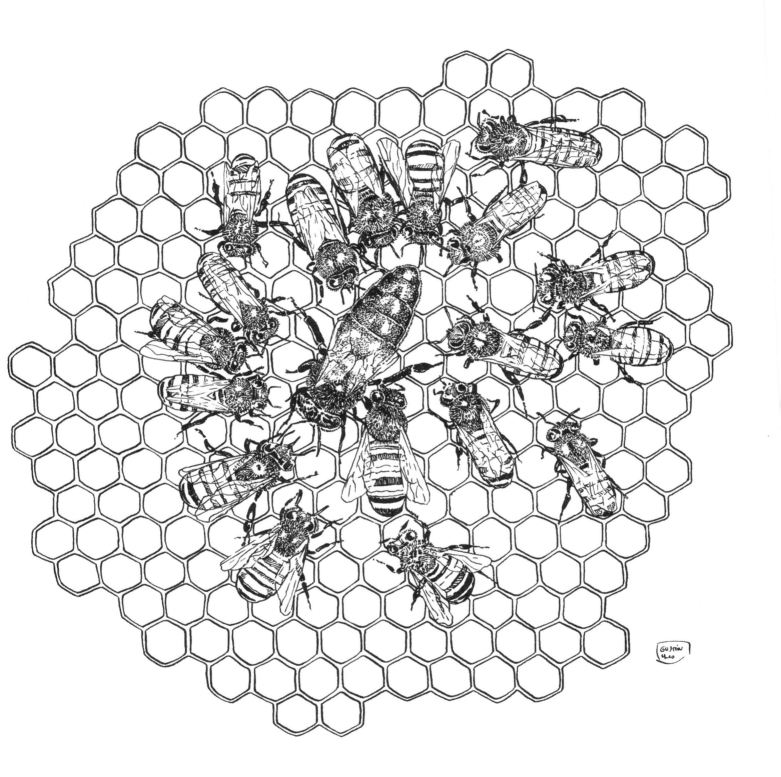

THE STING AND VENOM

DID YOU GET STUNG? IT COULDN'T HAVE HAPPENED AT A BETTER TIME (IF I MAY SAY SO), BECAUSE I WANTED TO TALK TO YOU ABOUT BEE VENOM. AS EVERYONE KNOWS, A STING IS ALWAYS PAINFUL, BUT REACTIONS DIFFER FROM ONE PERSON TO ANOTHER. THERE ARE PEOPLE WHO ARE IMMUNE, WHO SUFFER ONLY FROM A SLIGHT SWELLING WITH TINGLING, PEOPLE WHO HAVE DISCOMFORT, AND FINALLY ALLERGIC OR HYPERSENSITIVE PEOPLE WHO MUST TAKE GREAT PRECAUTIONS IN CASE OF STINGS.

HEY, FRIENDS! I'VE BECOME AN "APIPUNCTURIST"!

A little history to refresh your memory. It was during the 18th century that the Dutch naturalist Jean Swammerdan accurately gave the constitution of the sting. At that time, the technical means for studying insects were not what they are today.

IMMUNE

SENSITIVE

HYPERSENSITIVE

The properties of bee venom are such that there is a therapy called apipuncture today. This therapy, which seems new, is not. Already under Charlemagne, our ancestors the Francs had their feet and hands pricked to soothe their rheumatic pains. As the need for medicine and pharmacy has increased considerably, the various artisanal methods of collecting venom have been modernized, and electroshock is now used. (It takes about 10,000 bees to obtain just 0.035 oz. / 1 g of venom).

PLEASE! PLEASE! A LITTLE ATTENTION. BEFORE WE LOOK FURTHER AT THE SCHEMATICS OF OUR OFFENSIVE WEAPON, LET'S FIRST LOOK AT THE COMPOSITION AND PROPERTIES OF OUR VENOM.

AH ... THAT STINGING DID ME GOOD!

COMPOSITION
Many acids (formic, hydrochloric, phosphoric), histamine (hence its use in pharmacology), antibiotic substance, diastases, diffusion substance

PROPERTIES
ANTIRHEUMATIC
ANTICOAGULANT
ANALEPTIC
ANTINEURALGIC
VASODILATOR

THE SALE OF VENOM IS STRICTLY RESERVED FOR PHARMACEUTICAL PURPOSES.

"NOW THAT YOU HAVE STUNG MY HAND, LET ME EXPLAIN TO OUR READERS HOW YOUR DEFENSIVE SYSTEM WORKS."

The stinger penetrates slightly to the first barb, then by a reciprocating system sinks to the last barb (B). The latter acts as a lever and allows the bee to release itself, which causes the attacking device to be completely extracted. After that, the venom will inject itself completely if you don't take the precaution of immediately removing the stinger.

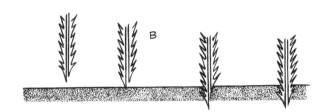

Never try to remove the sting by pinching the skin, since you would press on the venom sac, which would inject the poison. Using your fingernail or a knife, pull the stinger by grasping it from below. The sting comes out, along with the venom sac.

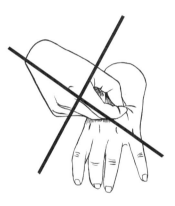

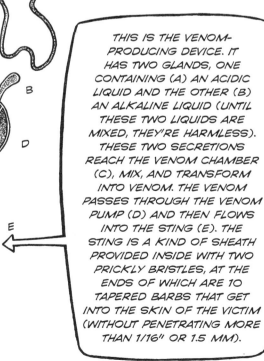

"THIS IS THE VENOM-PRODUCING DEVICE. IT HAS TWO GLANDS, ONE CONTAINING (A) AN ACIDIC LIQUID AND THE OTHER (B) AN ALKALINE LIQUID (UNTIL THESE TWO LIQUIDS ARE MIXED, THEY'RE HARMLESS). THESE TWO SECRETIONS REACH THE VENOM CHAMBER (C), MIX, AND TRANSFORM INTO VENOM. THE VENOM PASSES THROUGH THE VENOM PUMP (D) AND THEN FLOWS INTO THE STING (E). THE STING IS A KIND OF SHEATH PROVIDED INSIDE WITH TWO PRICKLY BRISTLES, AT THE ENDS OF WHICH ARE 10 TAPERED BARBS THAT GET INTO THE SKIN OF THE VICTIM (WITHOUT PENETRATING MORE THAN 1/16" OR 1.5 MM)."

"THE VENOM ENDS ITS JOURNEY BY BEING EXPELLED INTO THE WOUND..."

• THE STING AND VENOM •

"MOST BEEKEEPERS ARE IMMUNE, BUT FOR THOSE WHO AREN'T, PROTECTION IS NEEDED. IT SHOULD BE NOTED THAT SOME PARTS OF THE BODY ARE MORE VULNERABLE THAN OTHERS, SUCH AS THE FACE AND NECK. I ADVISE SENSITIVE PEOPLE TO WEAR APPROPRIATE CLOTHING (A), AND EVEN HARDENED AMATEURS TO KEEP A VEIL HANDY."

"TO AVOID GETTING STUNG TOO OFTEN, GO TO THE HIVE ON A SUNNY DAY. WIND AND THUNDERSTORMS MAKE BEES DEFENSIVE."

For allergic or hypersensitive people, there are desensitization centers where treatment helps with immunizing them against reactions. However, the results vary depending on the person.

For stings causing mild reactions, there are over-the-counter creams and ointments available.

"POPULAR REMEDIES ABOUND: RUBBING THE STING WITH PARSLEY, BASIL, ELDERBERRY OR NETTLE LEAVES, OR FRESH LEEK JUICE. AN EXCELLENT METHOD IS TO PLACE AN ICE CUBE ON THE STING."

"BE CAREFUL NOT TO CONFUSE THE STING OF A BEE WITH THAT OF A WASP OR HORNET, SINCE YOUR DOCTOR WILL ACT ACCORDINGLY. A BEE USUALLY LEAVES ITS STINGER AFTER IT STINGS YOU; A WASP NEVER DOES!"

• THE STING AND VENOM •

BEE LANGUAGE

A COLONY, CONSISTING OF BETWEEN 40,000 AND 60,000 INDIVIDUALS, OWES ITS GOOD FUNCTIONING TO THE DIFFERENT SENSES OF COMMUNICATION THAT BEES POSSESS. WE WILL BRIEFLY DISCUSS EACH OF THESE SENSES.

There are two antennae between the two large eyes and below the simple eyes. They have a diameter of 0.25 mm and are about 0.2 in. (5 mm) long. The antenna is divided into two parts: the scape (A), welded to the head, and the flagellum (B), with 10 sections for the female and 11 for the male, who has more sense of smell. The main functions of the antennas are to receive odors, communicate inside the hive, and transmit sensations. If we remove the antennas of a queen, bee, or male, they will be unable to devote themselves to their tasks. They will live but will no longer perform their functions in the hive: the queen will lay eggs anywhere and let herself go, the male will refuse food, and the bee will do nothing at all and will fly away without ever finding her way back.

THE TONGUE IS AN ORGAN USED TO RECOGNIZE THE ORIGIN OF NECTAR. INSIDE THE HIVE, IT HELPS BEES KNOW WHICH FLOWERS TO VISIT. IT'S CERTAIN THAT THE BEE HAS TASTE (IN THE GOOD SENSE OF THE WORD), BECAUSE IT KNOWS HOW TO DISTINGUISH BETWEEN BITTER AND SWEET FLAVORS. HOWEVER, THE MAIN PURPOSE OF THE TONGUE IS SUCKING UP LIQUIDS.

HOW DOES THE BEE INFORM ITS COMPANIONS OF A SOURCE RICH IN NECTAR? SHE DANCES! IT WAS PROFESSOR KARL VON FRISCH (1886–1982) WHO DISCOVERED THE MEANING OF BEE DANCING, WHICH WON HIM A NOBEL PRIZE IN 1973. LET'S LOOK AT THE BEHAVIOR OF A BEE THAT HAS JUST DISCOVERED A SOURCE OF SUPPLIES.

THE ROUND DANCE: AFTER FORAGING, THE BEE RETURNS TO THE HIVE TO HAVE DROPLETS OF THIS HONEY TASTED BY A GROUP OF BEES, WHICH IDENTIFIES THE NATURE OF THE FLOWER. IF THE NECTAR SOURCE IS LESS THAN 100 METERS AWAY, THE BEE STARTS TO MAKE SEMICIRCLES IN ONE DIRECTION AND IN ANOTHER. MANY BEES FOLLOW HER IN THIS DANCE, KEEPING IN TOUCH WITH HER THROUGH THEIR ANTENNAE. THE BEES THEN FLY STRAIGHT TO THE INDICATED SOURCE.

BACK IN THE HIVE, THEY WILL DO THE SAME DANCE TO OTHER BEES AND SO ON UNTIL THE NUMBER OF FORAGERS CORRESPONDS TO THE AREA TO BE EXPLOITED. WHEN THE HARVEST IS COMING TO AN END, THE BEES RETURNING TO THE HIVE NO LONGER PERFORM THIS DANCE, SO AS NOT TO SEND FORAGERS UNNECESSARILY.

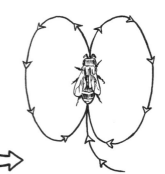
ROUND DANCE

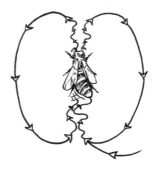
WAGGLE DANCE

THE WAGGLE DANCE: WHEN THE NECTAR SOURCE EXCEEDS 100 METERS, OUR BEE MAKES SEMICIRCLES IN ONE DIRECTION AND THEN IN ANOTHER (IN THE SHAPE OF AN EIGHT), AND WHEN SHE CROSSES THE STRAIGHT LINE OF THE TWO SEMICIRCLES, SHE MOVES HER ABDOMEN WITH A WIGGLE. THE GREATER THE DISTANCE BETWEEN THE HIVE AND THE FLOWER, THE FEWER DANCE REVOLUTIONS WILL BE PERFORMED IN A GIVEN TIME.

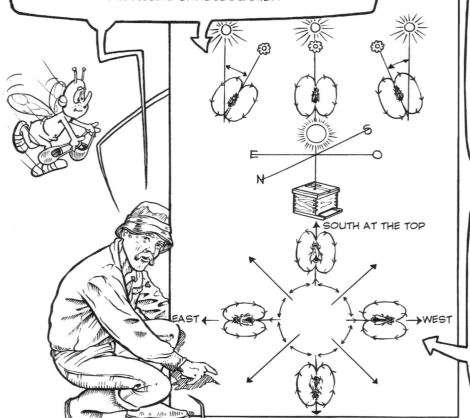

THE "DANCER" CAN PERFORM THESE DANCES ON THE BOTTOM BOARD, WALLS, OR FRAMES. SHE'LL DO ONE OR THE OTHER DANCE FOLLOWING THE CARDINAL POINTS, AND DEGREES RELATIVE TO THE SUN. IF THE FLOWER IS IN THE SOUTH, THE BEE IN ITS STRAIGHT LINE WILL WALK UPWARD; IN THE NORTH, IT WILL GO DOWN; IN THE WEST, IT WILL HEAD TO THE LEFT; IN THE EAST, IT WILL TURN TO THE RIGHT.

SO, WE BEES USE MEANS OF COMMUNICATION THAT HAVE INSPIRED SOME TECHNOLOGICAL PROGRESS ...

COLOR

THE BEE SEES ABOUT THE SAME COLORS AS WE DO, WITH THE DIFFERENCE THAT IT ALSO PERCEIVES ULTRAVIOLET RAYS. WHEN IT COMES OUT OF THE HIVE, IT ORIENTS ITSELF IN RELATION TO THE SUN.

THE COLORS TO WHICH IT IS MOST SENSITIVE ARE WHITE, BLACK, VIOLET, YELLOW, ORANGE, AND BLUE. THE COLOR OF YOUR HIVES AND THE GEOMETRIC SHAPE THAT MAY BE PAINTED ON THE FACADE OR BOTTOM BOARD ARE ALSO IMPORTANT IN ORIENTING THE BEES. A WORD OF ADVICE: AVOID USING GREEN AND RED, BECAUSE OUR LITTLE FRIENDS CONFUSE THESE TWO COLORS WITH BLACK; YOU COULD CAUSE A REAL MESS AT THE ENTRANCES. BUT COLOR IS NOT THE ONLY SOURCE OF ATTRACTION. IF BEES ARE FORAGING IN A CANOLA FIELD, IT'S NOT ONLY BECAUSE IT'S YELLOW, BUT ALSO BECAUSE OF THE SMELL THAT THE FLOWERS EMIT.

IF I WEAR THESE BEAUTIFUL COLORS, IT'S TO ATTRACT POLLINATING INSECTS. IN HIS TIME, DARWIN CARRIED OUT AN EXPERIMENT THAT SPEAKS FOR ITSELF: HE TOOK A LOBELIA ERINUS FLOWER AND DETACHED THE BLUE PETALS FROM THE COROLLA, AND THE FLOWER WAS NO LONGER VISITED.

Honey does not always have the same color. The soil, the flowers visited, but also the weather play an essential role in the coloring of the honey and, of course, its taste. A few years ago, each region had its own typical honey. Nowadays, intensive and standardized cultivation in all regions is tending toward the standardization of the color of honey—as desired by the consumer. Some beekeepers heat their honey to give it a more beautiful color or to remove sugar crystals. Why hide a natural phenomenon when it would be so simple to explain to consumers that these imperfections are due to the evolution of the product? These are glucose crystals that appear in a fructose and water solution.

DON'T TRY TO CHANGE MY COLOR. I'M LIVING FOOD.

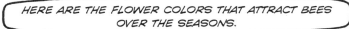

HERE ARE THE FLOWER COLORS THAT ATTRACT BEES OVER THE SEASONS.

SPRING	Flower color	Honey color
Acacia	White	Transparent
Hawthorn	White	Amber
Canola	Yellow	White
Eucalyptus	White	Amber
Alfalfa	Purple	Light yellow
Dandelion	Yellow	Yellow
Lime tree	Yellow	Pale yellow

SUMMER	Flower color	Honey color
Chestnut tree	Yellow	Dark brown
Lavender	Blue	Amber
Alfalfa	Purple	Light yellow
Dandelion	Yellow	Yellow
Sainfoin	Red	White
Fir tree	White	Dark brown
Sunflower	Yellow	Yellow
Blackberry	Pink	White

AUTUMN	Flower color	Honey color
Heather	Red	Reddish brown
Ivy	Greenish yellow	Clear
Blackberry	Pink	White
Thyme	Pink	Somewhat dark

WINTER	Flower color	Honey color
Heather	Red	Reddish brown
Dandelion	Yellow	Yellow

This season is not a good time for honey; bees are mainly looking for pollen.

I, THE POLLEN GRAIN, DON'T ALWAYS HAVE THE COLOR OF MY FLOWER. HERE ARE SOME EXAMPLES:

Plant	Pollen color	Flower color
Strawberry	Light brown	White
Chestnut	Red	White
Dandelion	Orange	Yellow
Willow	Gray	Yellow
Lime tree	Pale	Yellow
Borage	Green	Blue
Poppy	Black	Red
Sainfoin	Brown	Red
Heather	White	Red
Fir tree	Clear	Red
Hawthorn	Red	White

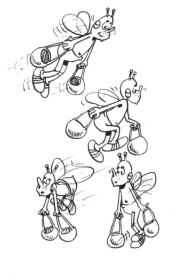

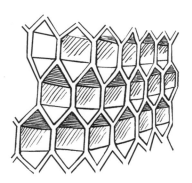

AT SECRETION, THE WAX IS CLEAR. WE USE IT TO BUILD CELLS. AFTER A FEW WEEKS, IT TURNS DARK YELLOW AND AGES INSIDE THE HIVE, WHERE IT BECOMES ALMOST BLACK.

TO MAKE IT EASIER TO SPOT US AND ALSO TO KNOW OUR AGE, BEEKEEPERS STICK A SMALL PATCH OF A DIFFERENT COLOR ON OUR BACK OR THORAX EACH YEAR, AN OPERATION CALLED QUEEN MARKING. SOME BEEKEEPERS PUT A COLORED PUSH PIN ON THE HIVE TO KNOW AT A GLANCE HOW OLD ITS QUEEN IS.

Red	Green	Blue	White	Yellow
—	—	2020	2021	2022
2023	2024	2025	2026	2027
2028	2029	2030	2031	2032

AND SO ON...

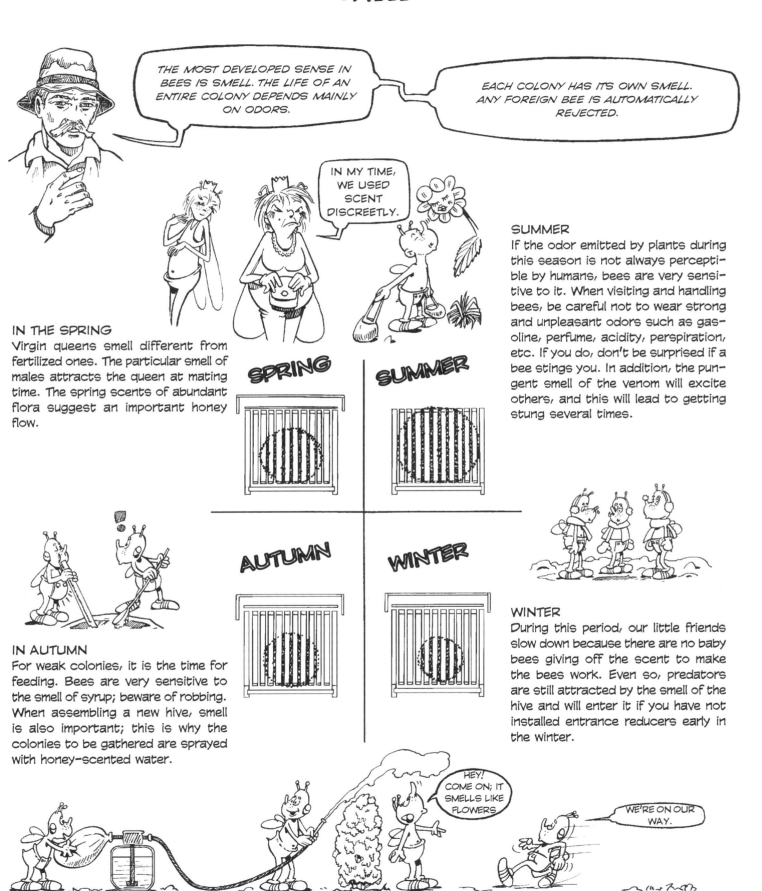

HEAT

 *NOW LET'S SEE WHAT THE CONSEQUENCES OF HEAT IN THE HIVE ARE OVER THE SEASONS.*

IN THE SPRING
As soon as the outside temperature reaches 50°F (10°C) the cluster begins to break up. When spring arrives, pollen abounds and the queen begins to lay eggs. The drones are born and help warm the brood before flying off to fertilize a queen. The population is growing. The volume of the colony expands as the temperature rises. It's time to put on the first super to avoid swarming.

SUMMER
With high temperatures, the internal temperature of the hive increases. Ventilation is added with shims, screened inner covers, or screened bottom boards; thanks to bees fanning their wings, the hive is ventilated and the atmosphere is cooled. It's time to add supers and later, to harvest.

SPRING

SUMMER

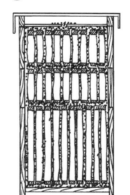

IN AUTUMN
The temperature and population are decreasing. The harvests are finished and the supers are removed, which makes the hive gain a few degrees. The drones are expelled and the young bees get ready for winter.

WINTER
The heat has disappeared! The cold has set in. If your hives are buried under the snow, don't worry; the cluster can withstand up to -31°F (-35°C). How do bees, you may ask, survive at such a low temperature? They produce energy by consuming and digesting honey. The substances resulting from this transformation are burned in contact with the oxygen that reaches the respiratory tract.

AUTUMN **WINTER**

THE TEMPERATURE INFLUENCES OUR FOOD CONSUMPTION. IF IT'S TOO COLD, WE EAT TO PRODUCE HEAT. IF IT'S TOO HOT, WE EAT TO COOL THE HIVE.

THE HIVE

BEFORE I TALK TO YOU ABOUT TODAY'S HIVES, THEIR FEATURES AND CONSTRUCTION, I'LL START WITH THE HIVES OF LONG AGO.

IN THE PAST, BEES USED TO LIVE IN HOLLOW TREES (AS WILD SWARMS DO NOWADAYS). HUMANS TOOK AN EXAMPLE FROM THEM AND SET UP THEIR OWN BEE COLONIES IN TREE TRUNKS. LATER, THE BEEKEEPER USED CLAY AND BRANCHES TO BUILD HIVES, THEN STRAW BECAME THE USUAL MATERIAL, AND FINALLY BOARDS (NOWADAYS, PLASTIC AND ALUMINUM WOULD TEND TO REPLACE THE PREVIOUS MATERIALS). SEE THE DIFFERENT MODELS OF TRADITIONAL HIVES, BELOW. MANY OTHERS WERE DESIGNED, BUT IN GENERAL THEIR DESIGN IS VERY SIMILAR TO THESE.

PRIMITIVE "LOG GUM" HIVES MADE FROM HOLLOW TREE TRUNKS

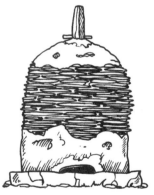

HORIZONTAL IN CORK   CLAY & BRANCHES SKEP   CLAY & BRANCHES SKEP   WICKER SKEP

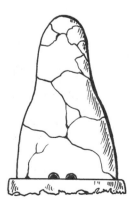

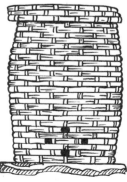
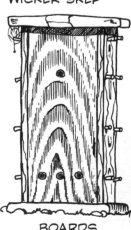

POTTERY   STRAW SKEP   STRAW   STRAW   BOARDS

THE HIVES YOU HAVE JUST SEEN ARE ONE PIECE AND MADE EITHER FROM A TREE TRUNK OR FROM WICKER, CLAY, BRANCHES, POTTERY, CORK, STRAW, AND EVEN FENNEL. THE HARVEST FROM THESE HIVES WAS VERY LIMITED, AND PARASITES COULD EASILY SETTLE IN. IN THE 1800S, THE HIVE WITH A SUPER (OR FALSE SUPER) WAS INVENTED. IN 1814, FRANÇOIS HUBERT INTRODUCED "MOVEABLE" FRAME HIVES. TOP BAR HIVES, FIRST USED BY EARLY EGYPTIANS, WERE INTRODUCED TO THE MODERN MARKET IN THE MID-1950S.

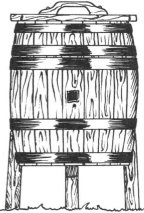
IN A BARREL

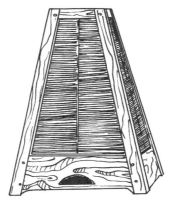
WOOD AND STRAW

STRAW

LOMBARD

GRAVENSHORT

STRAW WITH SUPER

STACKED

STRAW WITH SUPER

TOP BAR HIVE

RAVENEL   DELATRE

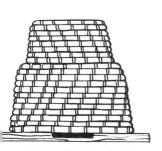
STRAW BODY AND SUPER

PALTEAU

BERLEPSCH

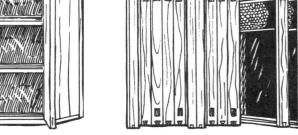
LEAF HIVE, BY F. HUBERT

• THE HIVE • 45

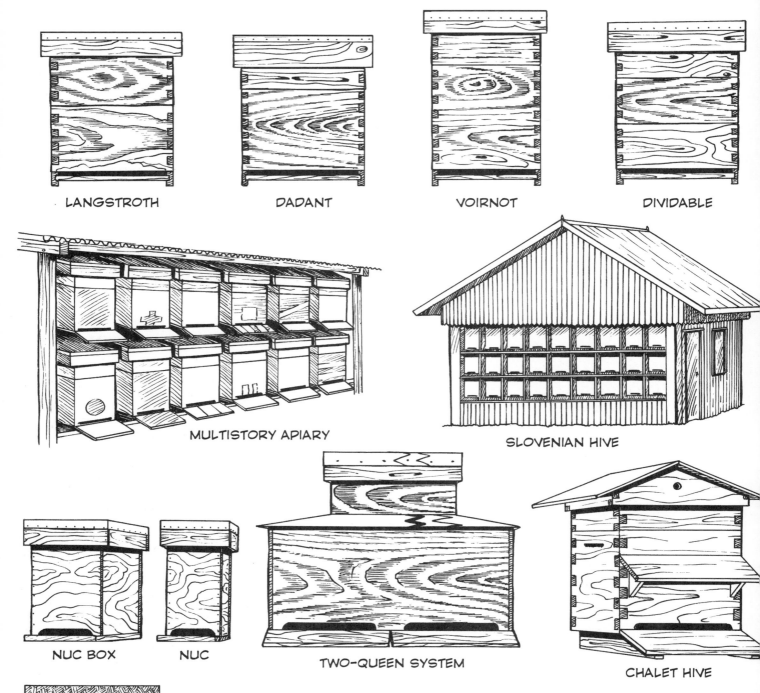
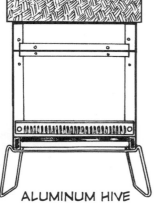

Here are the most common hives today, many found in Europe. The Langstroth consists of two identical bodies, the Dadant consists of a body and a super, the Voirnot because of its structure and its volume is similar to the tree trunk hive, and the dividable has supers placed one on top of the other (very practical for swarm making). The multistory apiary and the Slovenian hive provide a maximum number of sheltered hives in a minimum of space. The nuc box is used to receive swarms and for queen breeding. The two-queen system has two colonies, which sometimes allows a double harvest. The chalet hive is popular in cold regions, and the aluminum hive is well designed but less aesthetic.

46 • THE HIVE

HIVE CONSTRUCTION

BEFORE STARTING BEEKEEPING, MANY OF US DREAMED OF THE PERFECT HIVES WE SEE ON INSTAGRAM OR IN CATALOGS. BUT, FACED WITH THE DIFFICULTY OF BUILDING A HIVE, SOME HAVE GIVEN UP. LET ME OFFER YOU THE OPPORTUNITY TO BRING YOUR DREAM TO LIFE, BY REVEALING SOME OF OUR MANUFACTURING SECRETS.

I PRESENT TO YOU THE MOST COMMON MODELS: THE DADANT AND THE LANGSTROTH. THESE TWO TYPES OF HIVES ALLOW YOU TO FIND THE NECESSARY AND APPROPRIATE ACCESSORIES ON THE MARKET (SPACERS, REDUCERS, WAX FOUNDATIONS, ETC.).

The Dadant consists of 10 or 12 frames, depending on the desired model. It's very well adapted to hot and temperate regions. Plan three honey supers for strong colonies.

The Langstroth has identical hive body and honey super. Its volume is a little small for large populations. Some beekeepers replace the Langstroth honey supers with smaller honey supers.

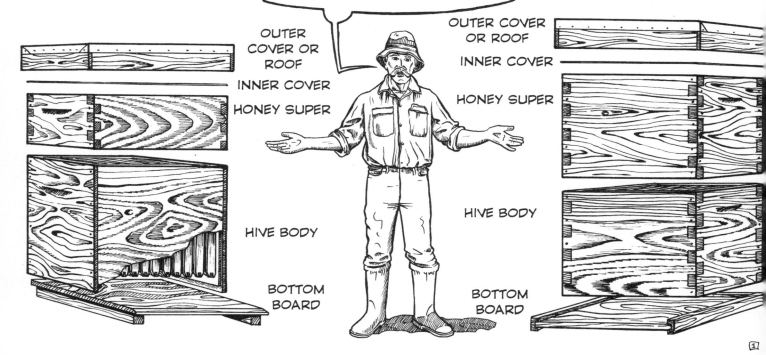

48 • HIVE CONSTRUCTION •

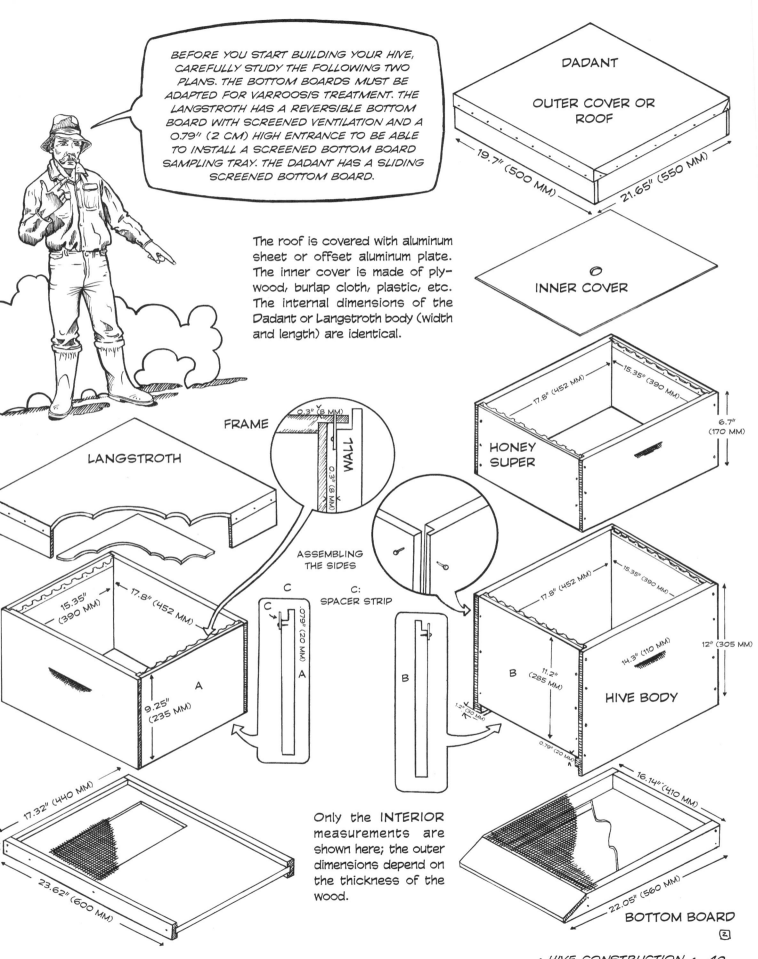

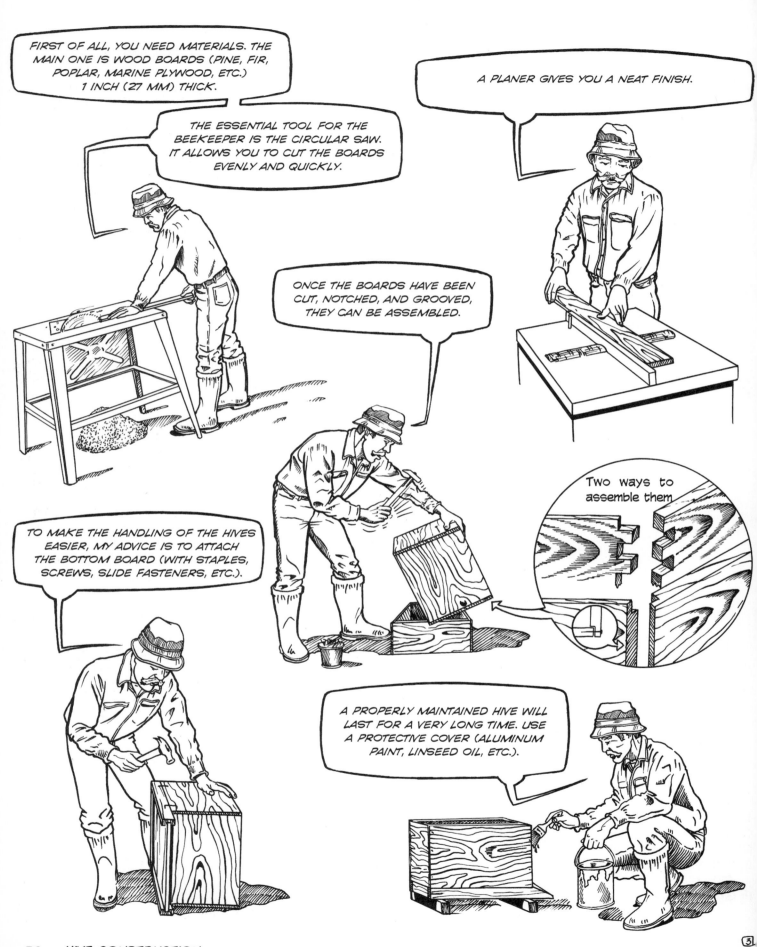

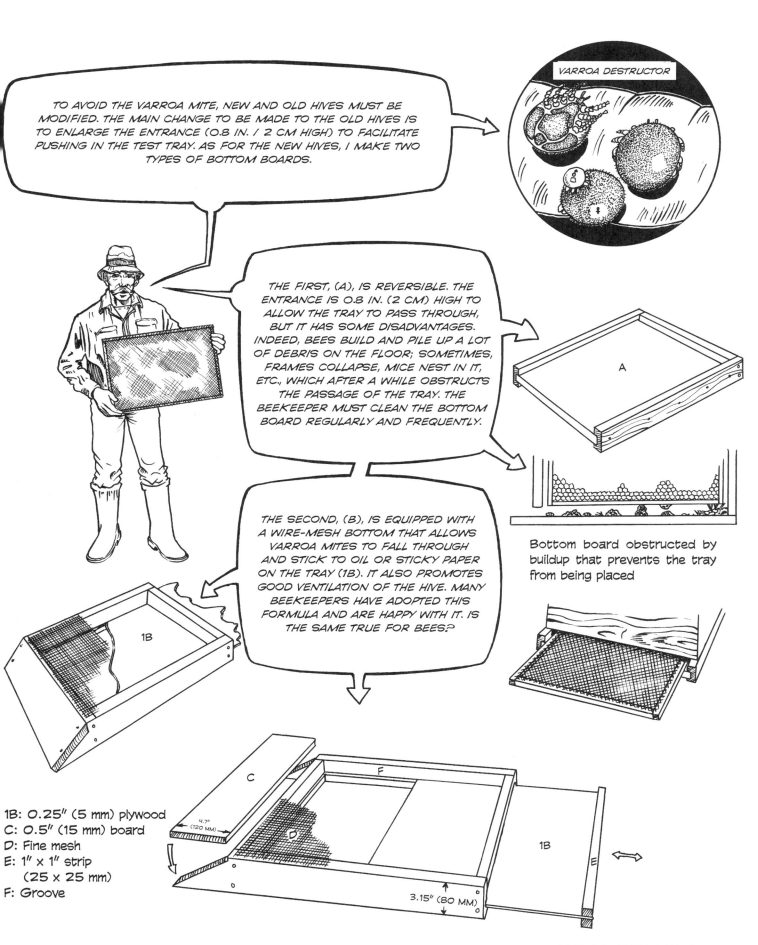

• HIVE CONSTRUCTION •

THE ACTIVITY OF A HIVE

LET'S NOW ENTER A BEEHIVE IN FULL ACTIVITY, HERE A MODEL DADANT. THERE ARE THREE PARTS:

—A BODY WITH 10 FRAMES, INCLUDING THE OUTSIDE FRAMES (CONTAINING HONEY AND POLLEN) AND THE CENTRAL FRAMES (FILLED WITH BROOD OF ALL AGES). THIS SITUATION REMAINS THE SAME ALL YEAR ROUND, EXCEPT DURING THE WINTER MONTHS WHEN BEES CLUSTER.
—A SUPER (TWO OR THREE FOR VERY STRONG COLONIES) WITH NINE FRAMES THAT WILL BE FILLED WITH HONEY AND POLLEN DURING THE VARIOUS NECTAR FLOWS
—A ROOF TO PROTECT THE HIVE FROM THE ELEMENTS AND UNDER WHICH A FEEDER CAN BE PLACED FOR THE WINTER (FEEDING IS DONE WHEN THERE'S NO HONEY SUPER ON).

1. Aluminum sheet
2. Roof
3. Inner cover
4. Feeder (placed here only for needs of the illustration). It can be replaced by bee candy in the winter.
5. Spider (some are good predators against wax moths)
6. Super placed for honey harvesting
7. Cell full of pollen
8. House bee depositing the nectar
9. Queen surrounded by her court
10. Male or drone. It warms the brood in early spring. Its main function is fertilization of the queen.
11. Hive wall
12. Drone cells
13. Eggs
14. Brood. The frames of the center are filled with it.
15. Worker bees building honeycomb
16. Queen cells
17. Festooning bees
18. Larvae
19. Water collector bee
20. Young guard bee
21. Ventilators
22. Guard
23. Forager
24. Entrance

LET'S EXAMINE THE ACTIVITY IN THE HIVE STEP BY STEP.

52 • THE ACTIVITY OF A HIVE •

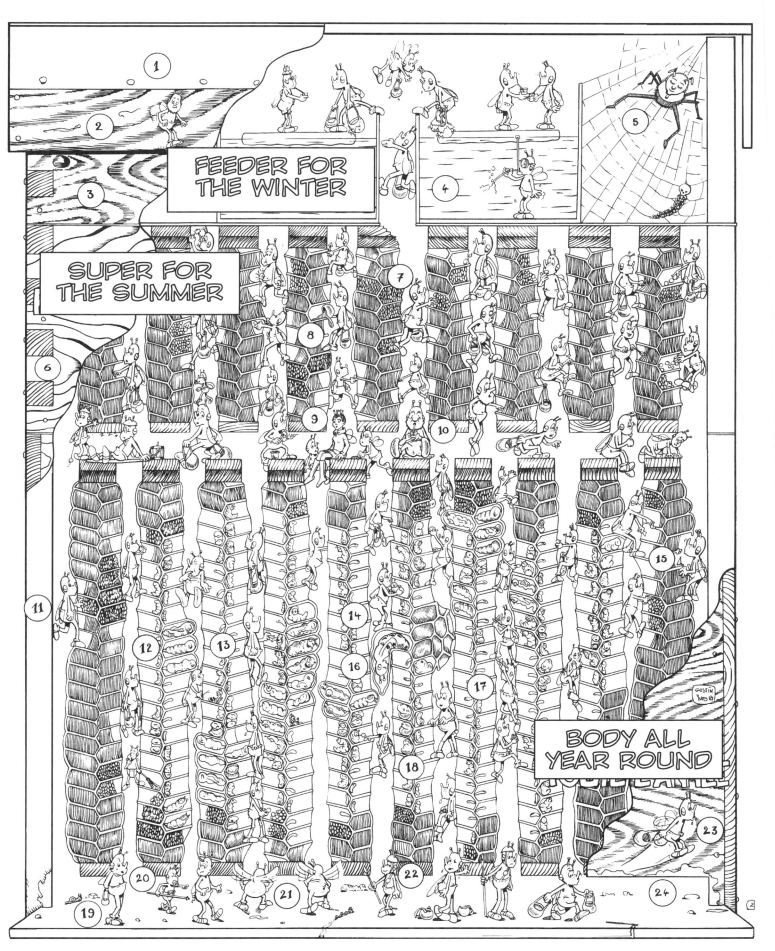

• THE ACTIVITY OF A HIVE •

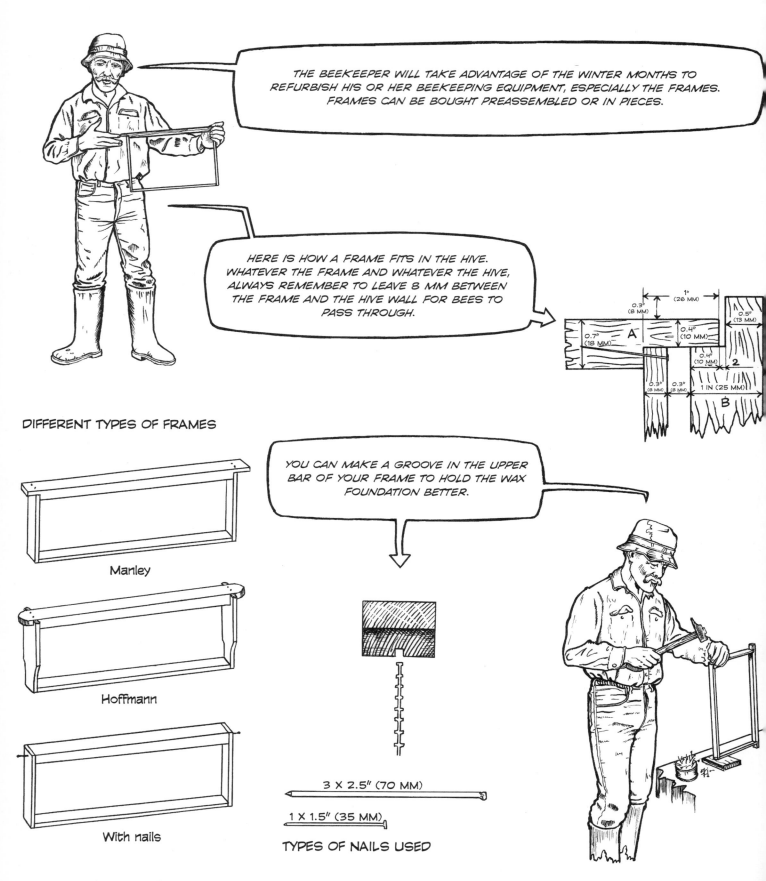

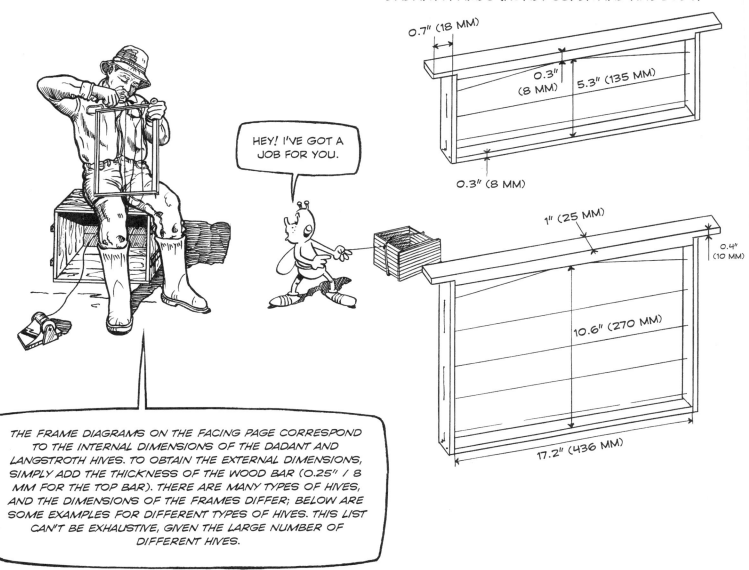

THE FRAME DIAGRAMS ON THE FACING PAGE CORRESPOND TO THE INTERNAL DIMENSIONS OF THE DADANT AND LANGSTROTH HIVES. TO OBTAIN THE EXTERNAL DIMENSIONS, SIMPLY ADD THE THICKNESS OF THE WOOD BAR (0.25" / 8 MM FOR THE TOP BAR). THERE ARE MANY TYPES OF HIVES, AND THE DIMENSIONS OF THE FRAMES DIFFER; BELOW ARE SOME EXAMPLES FOR DIFFERENT TYPES OF HIVES. THIS LIST CAN'T BE EXHAUSTIVE, GIVEN THE LARGE NUMBER OF DIFFERENT HIVES.

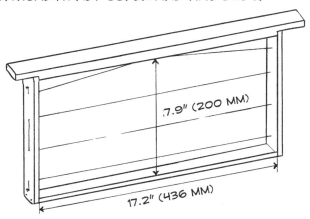

• BUILDING A FRAME • 55

INSTALLING WIRE

Although most beekeepers purchase frames that are already foundation-ready, whether wired or not, I prefer to wire my own frames, then embed the foundation sheets on the wire. If you're a DIY-type person, try it!

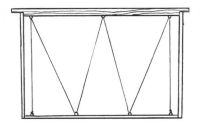  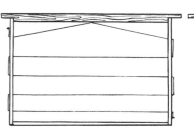 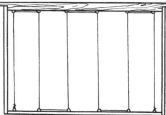

Different ways of arranging wire in the frames

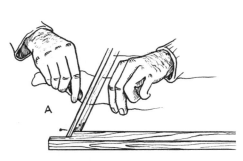

A: How to pass the wire through the drilled side bar by using a 0.10" / 2 mm diameter drill bit

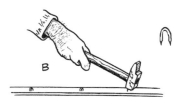

B: You can also pass the wire through small fence staples.

C: Ordinary staples can be used too.

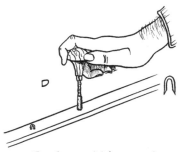

D: To avoid hammering your fingers, try a fence staple driver, found in hardware stores.

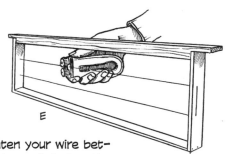

E: To tighten your wire better, use this small device: the zigzag wheel.

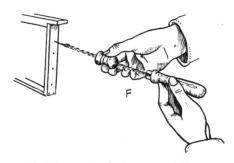

F: If you don't have a drill or a sophisticated device that drills five holes at a time, any drill will do.

EMBEDDING WAX FOUNDATION

IT'S COLD, IT'S RAINING ... IT'S A WINTER DAY AND THERE'S NO WAY WE'RE WORKING OUTSIDE. LET'S TAKE THE OPPORTUNITY TO INSTALL THE WAX FOUNDATION IN THE FRAMES. MAKE SURE THAT YOU'VE STORED YOUR FOUNDATION SHEETS CORRECTLY: FLAT, AND IN A DRY PLACE. (FRAMES WITH WAX FOUNDATIONS ARE STORED IN THE SAME WAY.) WHEN YOU BUY YOUR FOUNDATION, SEVERAL MATERIALS ARE AVAILABLE, INCLUDING PURE BEESWAX, MIXED WAX, AND PLASTICS.

MOST BEEKEEPERS PURCHASE FOUNDATION THAT'S READY TO PLACE IN FRAMES, BUT AS I MENTIONED ON THE PREVIOUS PAGES, I PREFER TO WIRE MY OWN FRAMES, THEN EMBED THE FOUNDATION SHEETS ON THE WIRE.

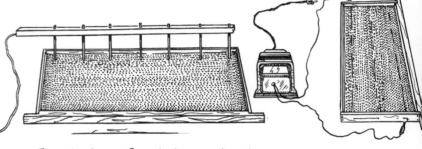

To embed your foundation on the wires, you can buy specialized equipment.

ARE YOU INTERESTED IN ANTIQUE BEEKEEPING METHODS OF EMBEDDING? THE LEROY RESISTANCE IS WORKED BY CUTTING ONE OF THE ELECTRIC WIRES AND PLACING AN OLD FLATIRON BETWEEN THE TWO PIECES. THE VINTUROUX MODEL CONSISTED OF A GLASS CONTAINER FILLED WITH 1 QUART (ABOUT 1 LITER) OF WATER AND A TABLESPOON OF COARSE SALT, ALL COVERED WITH AN INSULATING COVER (WOOD) PIERCED WITH TWO HOLES IN WHICH TWO NAILS WERE INSERTED.

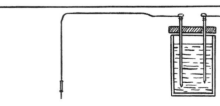

LEROY RESISTANCE

VINTUROUX RESISTANCE

I PREFER TO FILL MY FRAMES ENTIRELY WITH 100% BEESWAX FOUNDATION, SO I WAS ADVISED TO USE FOUNDATION IN ONLY CERTAIN FRAMES, AS STARTERS, SO TO SPEAK, TO SAVE MONEY. IN MY OPINION, THAT'S NOT WISE, BECAUSE BEES BUILD REGULAR, EVEN COMB AS LONG AS THEY'RE ON THE FOUNDATION, BUT BEYOND ITS LIMIT THEY TEND TO BUILD IRREGULAR CELLS, EVEN DRONE CELLS. THAT'S PROBLEMATIC FOR THE BEEKEEPER WHEN HANDLING THE HIVES. EVERYONE DECIDES WHICH METHOD IS BEST FOR THEM. PERSONALLY, I STICK WITH 100% WAX FOUNDATIONS ON EVERY FRAME.

PLACE A WOODEN BOARD THE SAME DIMENSIONS AS THE FOUNDATION SHEET TO APPLY PRESSURE. THIS WAY, THE WAX WILL INTEGRATE WELL WITH THE FRAME DURING THE SOLDERING PROCESS.

DRINKING WATER

Water is necessary and even essential for bees, especially in early spring, because nectar is not yet available in sufficient quantities. Bees need water for brood rearing.

A colony consumes about 1.3 gallons (5 liters) of water per month. If your apiary is near a field, be wary of the risk of water poisoning from ditches. Indeed, the chemicals used for crops pollute. Alas! The best way to provide water for your bees is to install a water trough in the apiary.

THE WORK IN THE APIARY OR WORKSHOP IS FINISHED, AND YOU'RE LOOKING FORWARD TO QUENCHING YOUR THIRST. WE NEED WATER TO LIVE; SO DO BEES!

MAKING A WATER TROUGH IS VERY SIMPLE. DEPENDING ON THE NUMBER OF COLONIES, YOU'LL USE AN UPENDED BOTTLE FROM WHICH THE WATER WILL DRIP, OR A CANISTER IN WHICH YOU WILL HAVE PLACED BRANCHES AND MOSSES TO PREVENT DROWNING. DON'T HESITATE TO PLACE THE WATER TROUGH IN THE SUN, BUT DO NOT EXPOSE IT TO THE WIND, SINCE BEES PREFER STAGNANT WATER TO FRESH WATER. IF YOU LOOK CAREFULLY AT THE BEES, YOU'LL SEE THAT THEY'RE LOOKING FOR WATER-CONTAINING ORGANIC DEPOSITS.

HERE'S A SIMPLE PROCESS FOR MAKING A WATER TROUGH: CUT OUT THE LID OF A CANISTER AND DRILL HOLES IN IT. NAIL TWO STRIPS UNDER THE LID TO ACT AS FLOATS. ONCE THE CANISTER IS FILLED WITH WATER, PUT THE FLOAT LID IN PLACE. IT WILL PREVENT THE BEES FROM DROWNING.

IF YOU DON'T HAVE A CAN, YOU CAN ALWAYS PROVIDE WATER TO YOUR BEES BY PLACING ANY CONTAINER FREE OF TOXIC MATERIALS IN YOUR APIARY. DON'T FORGET TO PUT TWIGS, MOSSES, BARK, ETC. ON THE SURFACE OF THE WATER.

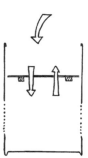

Lid

TO YOUR VERY GOOD HEALTH!

Using a chisel and a hammer, cut the lid in the same way as with a can opener, then clean your canister with sawdust and then with detergent. Rinse it well before filling it with clean water. Replace the drilled cover. The water trough is ready!

CLEANING THE APIARY

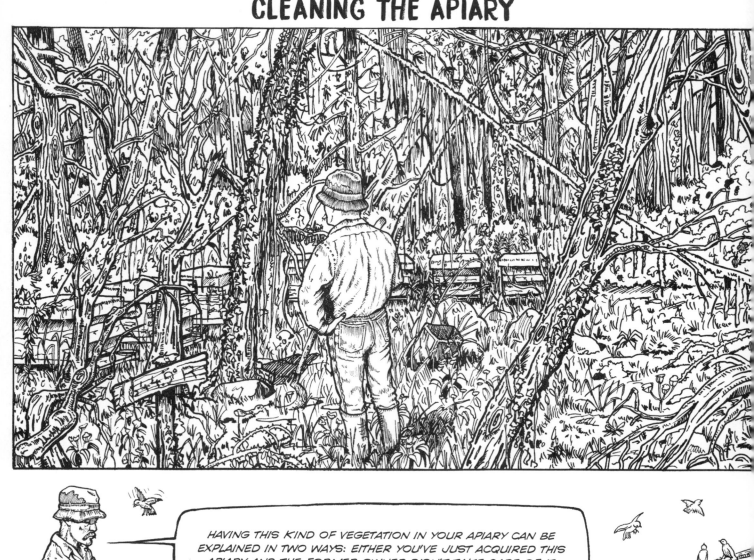

HAVING THIS KIND OF VEGETATION IN YOUR APIARY CAN BE EXPLAINED IN TWO WAYS: EITHER YOU'VE JUST ACQUIRED THIS APIARY AND THE FORMER OWNER DIDN'T TAKE CARE OF IT, OR YOU'VE BEEN AWAY TOO LONG.

CHOOSE A DRY, COLD, AND SUNNY AFTERNOON TO GET TO WORK. YOU'LL NEED DIFFERENT TOOLS: SCYTHE, SICKLE, BRUSH CUTTER, PRUNER, ETC.

PRUNING AND CLEANING WORK SHOULD BE CARRIED OUT WITHOUT SUDDEN MOVEMENTS, TO AVOID BRANCHES FALLING ON THE HIVES.

 THERE YOU GO ... THE JOB IS DONE, YOUR APIARY IS NEAT AND TIDY, AND YOU CAN SHOW IT OFF WITH PRIDE.

DURING YOUR FIRST SPRING VISITS, IF YOU HAVE ANY DOUBTS ABOUT THE HEALTH OF YOUR BEES, DON'T HESITATE TO CONTACT THE BEE INSPECTOR OR MASTER BEEKEEPER IN YOUR REGION.

THE FOLLOWING PAGES SHOW YOU SOME BEST PRACTICES FOR ARRANGING YOUR HIVES.

Don't be afraid to clear your hives on the side of the rising sun.

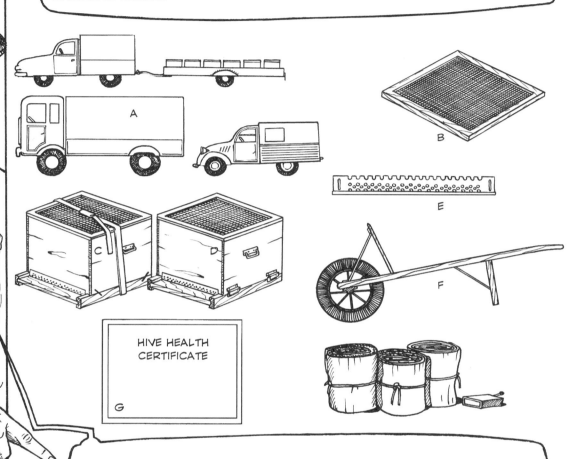

PREDATORS

PREDATORS AND PARASITES SHOULD NOT BE CONFUSED. AND PREDATORS SHOULD NOT BE SYSTEMATICALLY ELIMINATED, SINCE THEY'RE PART OF THE NATURAL CYCLE OF LIFE.

ANTS

A high concentration of ants can become a nuisance to the hive when they use an inner cover as an artificial incubator to hatch their eggs. It's then almost impossible to dislodge them: no matter how much you brush them off, the next day they're back again! No effective, harmless-to-bees treatment exists to prevent them from entering the hive. However, on plants, aphids are useful. They gorge themselves on sap, and their excretions fall on the leaves; the bees harvest this "honeydew."

WASPS

This insect ignores danger, makes its nest anywhere, and eats everything it can! Like the hornet, the wasp attacks its prey in flight. Its sting is becoming more and more dangerous, so you have to be very careful. In spring, too many of them make things difficult for beekeepers. A simple and effective way to destroy wasps and hornets is to place plastic bottles cut in half, with the upper part turned upside down. Pour inside some beer or wine as bait. (Never use honey in wasp traps; it will attract and kill honeybees.) Place your traps in areas such as honey houses, or hang them under a tree. The number of wasps and hornets that are caught will surprise you!

HORNETS

Its sting is particularly painful and sometimes fatal. The hornet's stinger is not barbed and does not remain in the epidermis. This insect is carnivorous and a scavenger. When it has nothing left to eat, it catches bees in flight. Frequently, hornets build their nests in inner covers and bait hives. Always lift the roofs gently. To get rid of this insect, the queen hornet should be destroyed in early spring.

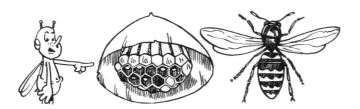

74 • PREDATORS •

MICE AND VOLES

When they enter the hive, these small rodents cause a lot of damage. They build their nest between the frames by using straw and dry leaves, damaging the wax and honeycomb. To avoid their intrusion, place entrance reducers or mouse guards at the beginning of winter.

SPIDERS

Most spiders aren't dangerous to bees. However, there are a few deadly species, which don't hesitate to catch bees, wasps, hornets, butterflies, etc. in their web. By carefully observing the behavior of spiders—most often in the outer cover—you'll easily recognize those that block the entry of unwanted insects, such as the greater wax moth, and should therefore be left alone.

THE GREATER WAX MOTH

A weak or sick colony does not escape this predator. Just a single one of these moths entering a hive is able to destroy a weak colony, since it can lay 200 eggs. It's not the moth itself that causes the damage, but the larvae that dig tunnels in wax and wood. Stored supers aren't immune to this predator.

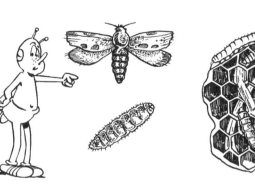

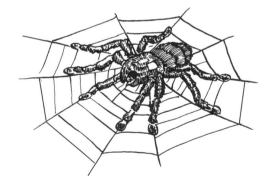

HUMANS

Humankind is certainly the greatest predator. By manufacturing and using pesticides and other toxic products, humans pollute and destroy the natural environment: fauna, flora, rivers, lakes, air. Sometimes vandals shoot hives, knock them down, and even use insecticides to kill the bees within.

• PREDATORS • 75

SOME NECTAR-PRODUCING PLANTS TO GROW

L = *Latin name*
F = Family
FS = Flowering season

BORAGE
L: *Borrago officinalis*
F: Borraginaceae
FS: May to September

HEATHER
L: *Calluna vulgaris*
F: Ericaceae
FS: June to September

THISTLE
L: *Cirsium lanceolatum et arven*
F: Asteraceae
FS: June to September

CANOLA
L: *Brassica napus*
F: Cruciferous
FS: April to May

LAVENDER
L: *Lavandula officinalis*
F: Lamiaceae
FS: July to August

IVY
L: *Hedera helix*
F: Araliaceae
FS: September to October

DANDELION
L: *Taraxacum officinale*
F: Asteraceae
FS: March to November

ALFALFA
L: *Medicago sativa*
F: Papilionaceae
FS: May to September

BLACKBERRY
L: *Rubus fruticosus*
F: Rosaceae
FS: May to August

SAINFOIN
L: *Onobrychis sativa*
F: Papilionaceae
FS: May to July

THYME
L: *Thymus vulgaris*
F: Labiaceae
FS: May to October

SUNFLOWER
L: *Helianthus annuus*
F: Asteraceae
FS: July to August

SOME SOURCES OF VARIETAL HONEY FLAVORS

HAWTHORN
L: *Crataegus monogyna*
F: Rosaceae
FS: April to May

BOXWOOD
L: *Buxus sempervirens*
F: Buxaceae
FS: March to April

OAK
L: *Quercus* spp.
F: Fagaceae
FS: April to May

MAPLE
L: *Acer*
F: Acetaceae
FS: April to May

HOLLY
L: *Ilex aquifolium*
F: Aquifoliaceae
FS: May to June

LOCUST
L: *Robinia pseudacacia*
F: Papilionaceae
FS: May to June

FIR, SPRUCE, PINE, ETC.
Excretions of aphids are harvested by bees in summer

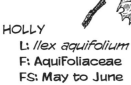

WILLOW
L: *Salix alba*
F: Salicaceae
FS: April to May

ELDERBERRY
L: *Sambucus nigras*
F: Caprifoliaceae
FS: June

LINDEN
L: *Tilia cordata*
F: Malvaceae
FS: June to July

• SOURCES OF VARIETAL FLAVORS • 77

BENEFICIAL SPECIES FOR THE APIARY

HAWTHORN
Crataegus oxyacantha

LOCUST
Robinia pseudacacia

BOXWOOD
Buxus sempervirens

PUSSY WILLOW
Salix caprea

LINDEN
Tilia cordata

SYCAMORE MAPLE
Acer pseudoplatanus

ASH
Fracinus excelsior

DYER'S BROOM
Genista tinctoria

COMMON CHESTNUT
Castanea vulgaris

GORSE
Ulex europaeus

OAK
Quercus robur

78 • BENEFICIAL SPECIES •

PEAR
Pirus communis

WILD CHERRY
Prunus avium

HAZEL
Corylus avellana

APPLE
Malus communis

WHITE BIRCH
Betula alba

DOMESTIC ROWAN
Sorbus domestica

WILD PLUM
Prunus americana

HORSE CHESTNUT
Aesculus hippocastanum

HOLLY
Ilex aquifolium

BLACKBERRY
Rubus fruticosus

IVY
Hedera helix

• BENEFICIAL SPECIES •

GRAFTING (FOR MORE TREES!)

CLEFT GRAFTING

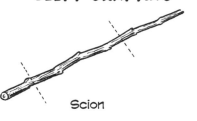
Scion

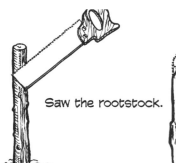
Saw the rootstock.

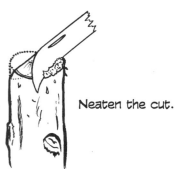
Neaten the cut.

Split the rootstock.

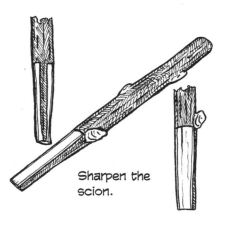
Sharpen the scion.

Drive in the scion.

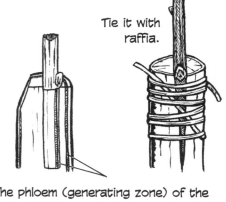
Tie it with raffia.

The phloem (generating zone) of the scion and the rootstock must coincide.

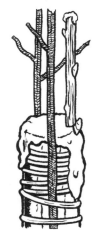
Seal the cleft with cold putty and protect the scion with twigs.

SHIELD GRAFTING

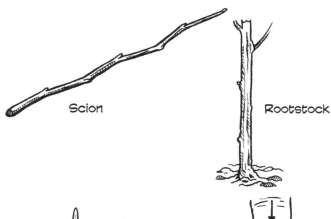
Scion — Rootstock — Make a T-shaped incision. — Loosen the bark.

Remove an oval of the stem (including the petiole and the bud).

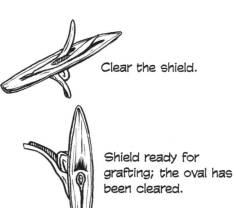
Clear the shield.

Shield ready for grafting; the oval has been cleared.

Place the shield into the T on the wood stock with your fingers, press with the back of the knife, and slice the graft.

The graft is complete.

Tie it with raffia.

INSTALLING HONEY SUPERS

WHEN IS THE RIGHT TIME TO INSTALL THE HONEY SUPERS? IT CAN'T BE TOO SOON OR TOO LATE! A REAL DILEMMA? IF YOU VISIT YOUR HIVES REGULARLY, YOU'LL EASILY JUDGE WHEN IT'S APPROPRIATE TO INSTALL THE FIRST HONEY SUPER. A STRONG BEE POPULATION IS A FIRST INDICATION, AND IF FLOWERING IS ABUNDANT, THE TIME HAS COME. IT'S BETTER TO PLACE THE HONEY SUPER TOO EARLY RATHER THAN TOO LATE, TO AVOID POSSIBLE SWARMING. SOME BEEKEEPERS, DUE TO LACK OF TIME OR SPACE FOR STORING THEIR SUPERS, LEAVE THEM ON THE HIVE BODY ALL WINTER, REMOVING THEM ONLY AT HARVEST TIME.

IF THE HIVE BODY FRAMES ARE "SWOLLEN" WITH WHITE WAX, DON'T HESITATE ANY LONGER!

IF YOU'RE WORRIED THAT THE QUEEN WILL LAY EGGS IN THE HONEY SUPER, INSERT A QUEEN EXCLUDER BETWEEN THE HIVE BODY AND THE HONEY SUPER. BEE SUPPLY RETAILERS OFFER A WIDE CHOICE OF EXCLUDERS.

IT'S SAD TO FIND BROOD IN THE SUPER FRAMES!

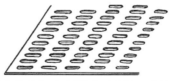

PLASTIC QUEEN EXCLUDER

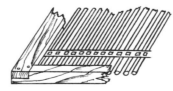

GALVANIZED QUEEN EXCLUDER MOUNTED ON A WOODEN FRAME

IF YOU'RE FORCED TO PLACE THE HONEY SUPER A LITTLE TOO EARLY, TAKE THE PRECAUTION OF SLIPPING A NEWSPAPER SHEET BETWEEN THE SUPER AND THE BODY TO AVOID COOLING THE BROOD NEST.

WHEN THE BEES NEED ACCESS TO THE HONEY STORE TO DEPOSIT THEIR LOOT, THEY WILL GRADUALLY TEAR UP THE NEWSPAPER.

A NEWSPAPER ON THE CEILING? NOW WE'VE SEEN EVERYTHING!

WHEN YOU ENTER THE HIVE, DON'T FORGET TO READ THE NEWSPAPER, UP THERE!

IF YOU WANT TO COLLECT PROPOLIS IN ADDITION TO HONEY, PLACE A PROPOLIS TRAP ON THE TOP OF THE HIVE. IT'S A PERFORATED PLATE MADE OF SOFT PLASTIC. THE BEES WILL HASTEN TO FILL UP ALL THE HOLES. WHEN THE TRAP IS FULL, REMOVE IT AND PLACE IT IN THE FREEZER TO HARDEN THE PROPOLIS. THEN ALL YOU HAVE TO DO IS ROLL UP YOUR TRAP SO THAT THE PROPOLIS COMES OFF, AND YOU GET A PERFECTLY CLEAN PRODUCT.

IN PRINCIPLE, A HONEY SUPER INCLUDES NINE FRAMES. BUT IF YOU WANT TO OBTAIN LARGER (I.E., THICKER) COMBS TO COLLECT MORE WAX, INSTALL ONLY EIGHT FRAMES. DON'T FORGET WHEN INSTALLING YOUR FRAMES TO PLACE THE CORRESPONDING SPACING STRIPS.

• INSTALLING HONEY SUPERS • 83

HARVESTING

HARVEST TIME HAS COME. IT IS BEST TO DO THIS ON A SUNNY DAY, WITHOUT THE THREAT OF A STORM. DON'T GET ANY ILLUSIONS, MY FRIENDS; NO MATTER HOW CAREFULLY YOU TAKE THE NECESSARY PRECAUTIONS TO REMOVE THE HONEY SUPERS, YOU'LL STILL GET STUNG BY SOME BEES WHO WON'T APPRECIATE YOU APPROPRIATING THE FRUIT OF THEIR EFFORTS.

BEFORE REMOVING THE HONEY SUPERS, CHECK THAT THEIR FRAMES ARE CAPPED.

YOU CAN REMOVE THE HONEY SUPERS IN DIFFERENT WAYS.

1. USING A BEE ESCAPE. IT'S AFFIXED TO THE INNER COVER AND ALLOWS THE BEES TO GO DOWN BUT PREVENTS THEM FROM COMING BACK UP INTO THE SUPER. IT'S A ONE WAY DOOR.

2. WITH A FUME BOARD PLACED ON THE SUPER. SINCE THEY DON'T LIKE THE SMELL, THE BEES ALMOST INSTANTLY DESCEND INTO THE HIVE BODY.

ROUND PLASTIC BEE ESCAPE

CONE BEE ESCAPE

PORTER BEE ESCAPE

STORING SUPERS

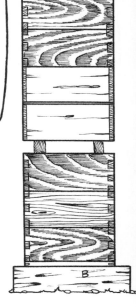

"Whether the harvest was good or bad, the process of storing the supers is the same."

"Before storing the extras, you will have taken the precaution of having them clean and dry. This operation consists of having the frames cleaned by the bees.
1st possibility (A): Place the super back on the hive.
2nd possibility (B): Pile your supers on the hives, remembering to insert entrance reducers to ease the entry and exit of the bees, and put on a lid to protect them from the rain."

"With the cleaning done, let's move on to storage: here again, several methods are possible. Either leave the honey supers on the hive or stack them."

"Remember to remove frames containing pollen, since the greater wax moth is fond of it. In spring, you'd be unpleasantly surprised to discover empty wax frames swarming with wax moth larvae."

90 • STORING SUPERS •

· STORING SUPERS · 91

"Once extracted, should the honey be heated? Each beekeeper has his or her own way of doing things. For my part, I find this process unnatural."

"I love good honey."

"Honey contains diastases useful for digestion. If it is overheated, these enzymes are destroyed, and, since it's no longer a living food, honey loses its virtues."

Attention: There are different heating practices! (1) Pasteurization: This is carried out by heating the honey to a temperature ranging from 158°F to 176°F (70°C–80°C) for a specific time. This technique deters the crystallization and fermentation of honey; it thus makes it possible to stabilize it and give it a marketable appearance! (2) Heating the honey to about 95°F (35°C): This temperature allows it to be put into jars; it's not harmful because it corresponds to the temperature of the hive.

"You can heat your honey in different ways without altering it. The hobbyist will use the double-boiler method. The semiprofessional will find honey heaters adapted to his or her needs. The professional will have the choice of various tools found at bee supply retailers: heating belts for drums, ovens for four or six drums, and hot rooms for more than six drums."

"If you want to give the consumer high quality honey, it should be extracted only when it is capped. Store it in a cool but not damp place. And above all, take the time to explain to the buyer that crystallized honey is perfectly natural."

• HONEY PROCESSING • 97

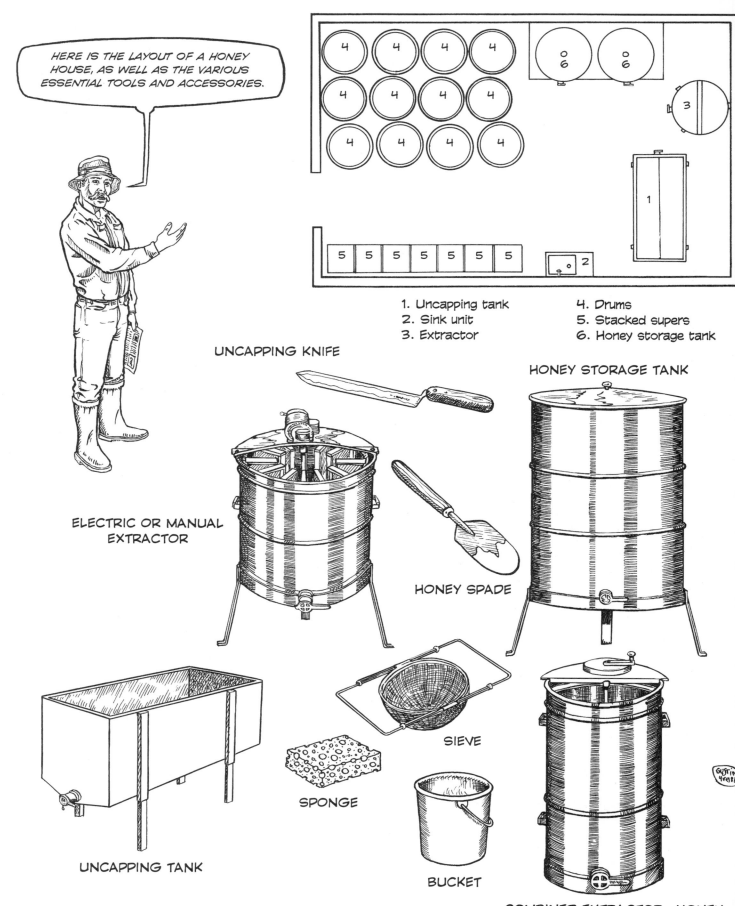

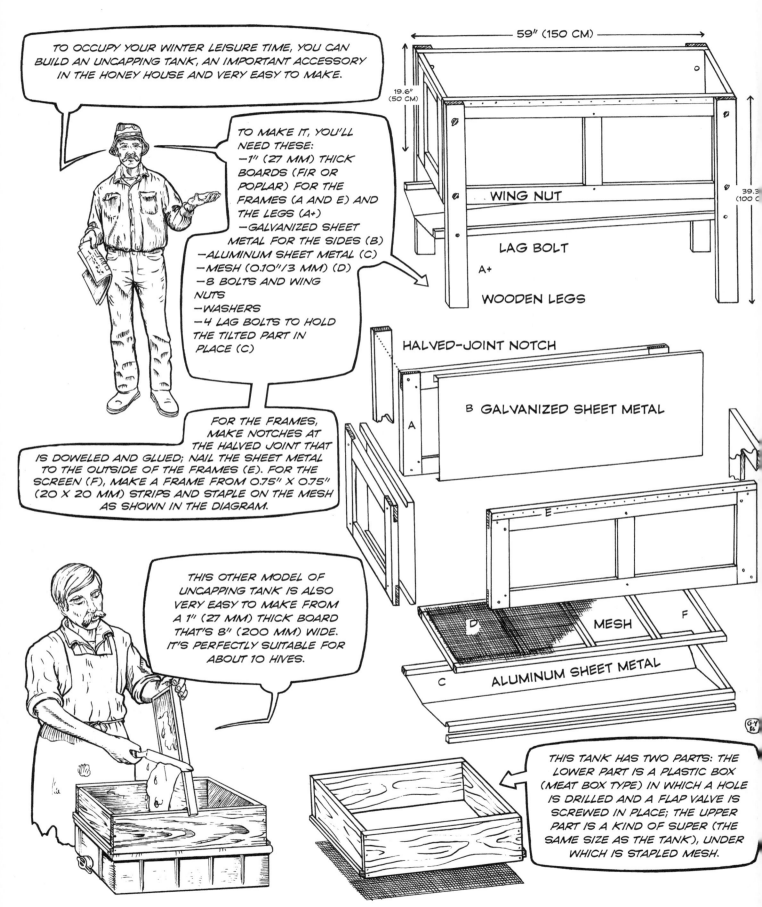

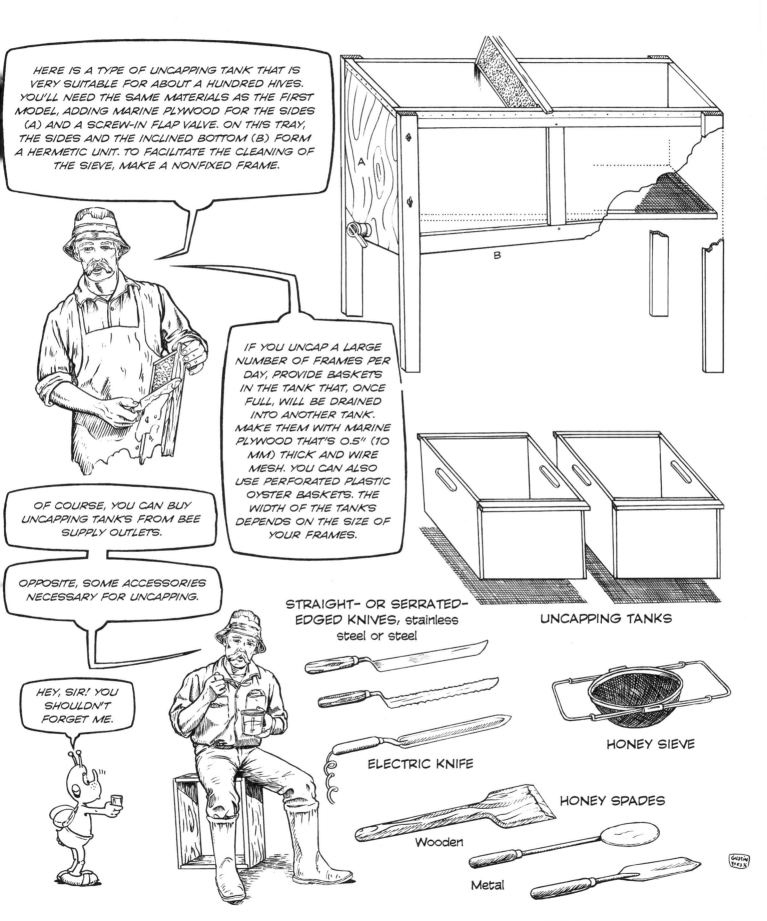

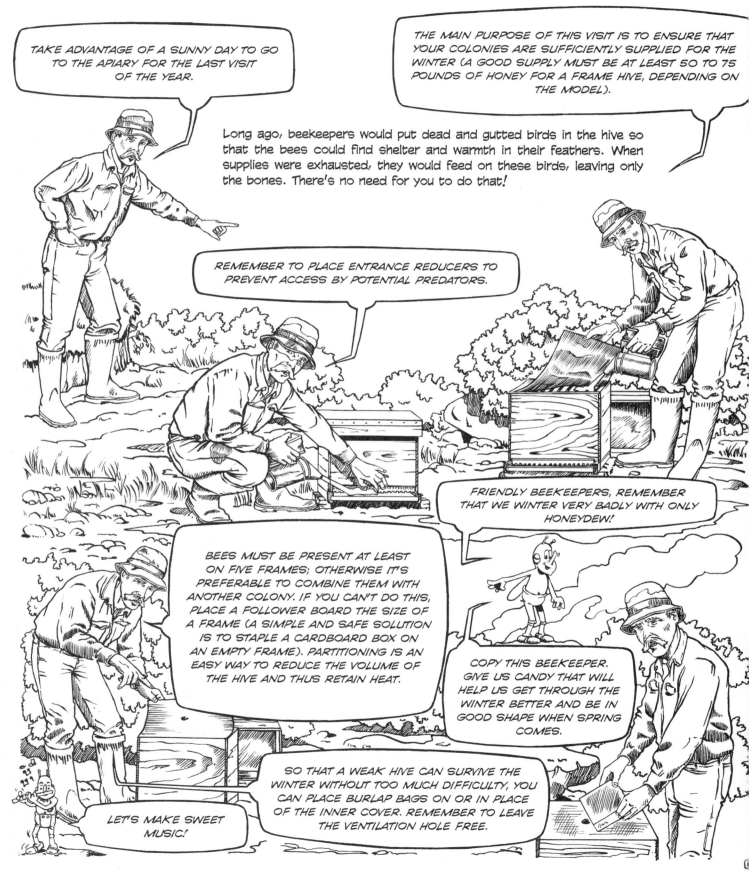

MAKING BEE CANDY

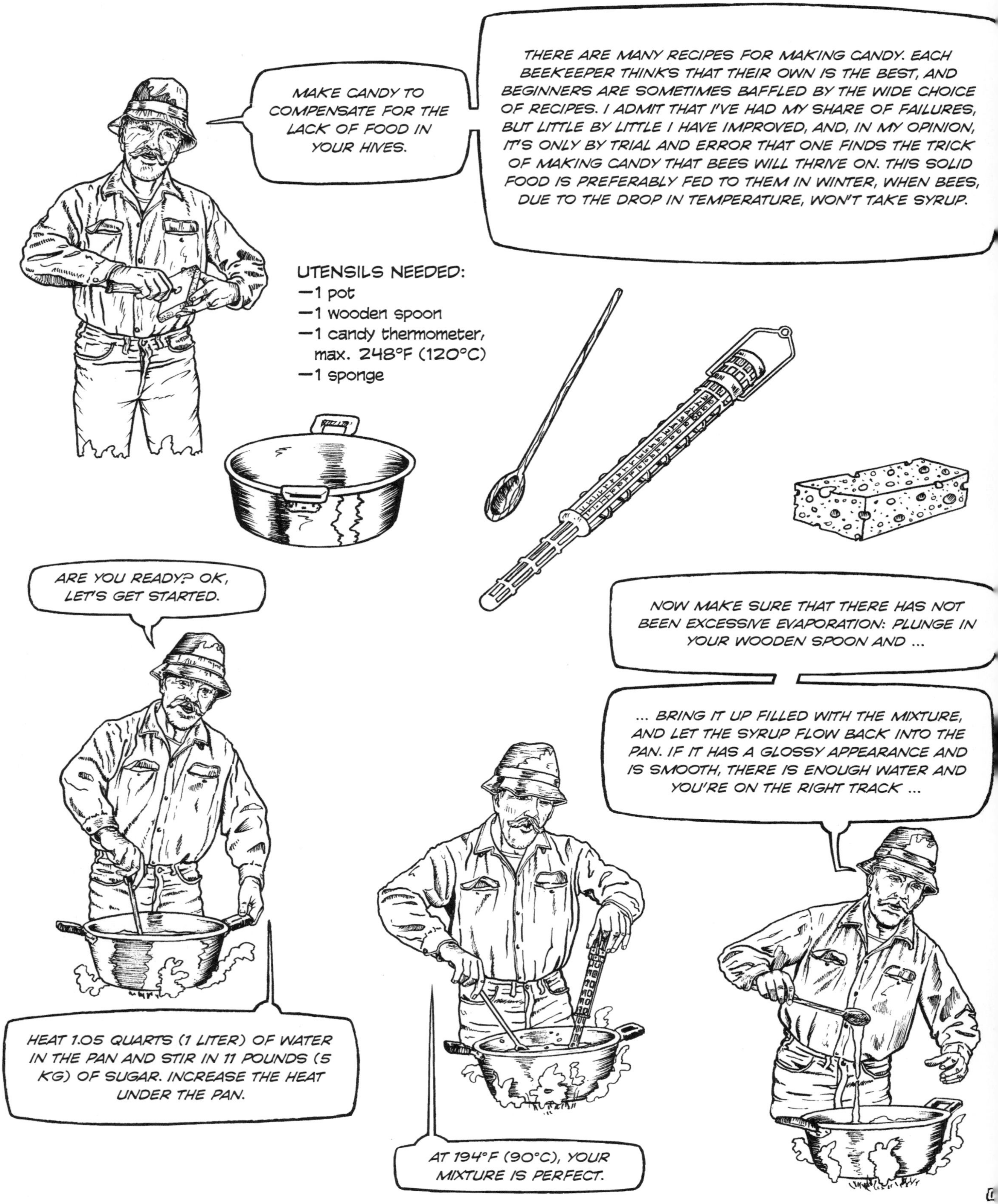

MAKE CANDY TO COMPENSATE FOR THE LACK OF FOOD IN YOUR HIVES.

THERE ARE MANY RECIPES FOR MAKING CANDY. EACH BEEKEEPER THINKS THAT THEIR OWN IS THE BEST, AND BEGINNERS ARE SOMETIMES BAFFLED BY THE WIDE CHOICE OF RECIPES. I ADMIT THAT I'VE HAD MY SHARE OF FAILURES, BUT LITTLE BY LITTLE I HAVE IMPROVED, AND, IN MY OPINION, IT'S ONLY BY TRIAL AND ERROR THAT ONE FINDS THE TRICK OF MAKING CANDY THAT BEES WILL THRIVE ON. THIS SOLID FOOD IS PREFERABLY FED TO THEM IN WINTER, WHEN BEES, DUE TO THE DROP IN TEMPERATURE, WON'T TAKE SYRUP.

UTENSILS NEEDED:
— 1 pot
— 1 wooden spoon
— 1 candy thermometer, max. 248°F (120°C)
— 1 sponge

ARE YOU READY? OK, LET'S GET STARTED.

NOW MAKE SURE THAT THERE HAS NOT BEEN EXCESSIVE EVAPORATION: PLUNGE IN YOUR WOODEN SPOON AND ...

... BRING IT UP FILLED WITH THE MIXTURE, AND LET THE SYRUP FLOW BACK INTO THE PAN. IF IT HAS A GLOSSY APPEARANCE AND IS SMOOTH, THERE IS ENOUGH WATER AND YOU'RE ON THE RIGHT TRACK ...

HEAT 1.05 QUARTS (1 LITER) OF WATER IN THE PAN AND STIR IN 11 POUNDS (5 KG) OF SUGAR. INCREASE THE HEAT UNDER THE PAN.

AT 194°F (90°C), YOUR MIXTURE IS PERFECT.

"DON'T DO LIKE I DID AND LET IT BOIL OVER!"

BE CAREFUL! THE TEMPERATURE RISES QUICKLY, AND AT 230°F (110°C) THE SYRUP FOAMS AND MAY BOIL OVER. TO AVOID THAT, DAMPEN YOUR SPONGE AND RUN IT OVER THE INNER WALL OF THE PAN UNTIL YOU TOUCH THE LIQUID. AT THAT MOMENT, BOILING STABILIZES. HEAT UP TO 242°F (117°C). IF POSSIBLE, HOOK YOUR THERMOMETER TO THE PAN SO THAT IT LEAVES YOUR HANDS FREE. TURN OFF THE HEAT AS SOON AS THE TEMPERATURE REACHES 244°F (118°C); OTHERWISE YOU'LL RUIN YOUR CANDY.

NEXT, ADD 2.2 POUNDS (1 KG) OF HONEY THAT YOU HAVE PREVIOUSLY HEATED IN A DOUBLE BOILER. AS IT COOLS, DON'T STIR THE MIXTURE. WHEN THE COOLING IS ...

... ENOUGH, USING YOUR WOODEN SPOON, STIR VIGOROUSLY UNTIL THE WHOLE THING LOSES ITS TRANSPARENCY AND BECOMES OPAQUE AND WHITISH. DON'T STOP AT THIS STAGE, OR YOUR CANDY WILL HARDEN. POUR IT INTO YOUR MOLDS.

"MMM! BEE CANDY."

"POUR INTO THE MOLDS WITHOUT DELAY."

• MAKING BEE CANDY •

"DON'T WASTE MONEY ON YOUR MOLDS; USE OLD ALUMINUM CONTAINERS OR MILK CARTONS (TETRA CARTON PACKAGING)."

A useful bee candy mold: a carton of milk cut in half lengthwise.

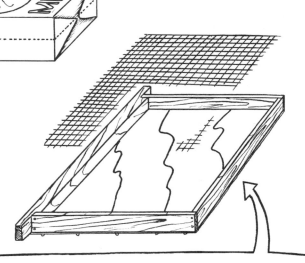

"YOU CAN ALSO POUR YOUR CANDY INTO A FRAME THAT YOU'VE SEALED UP ON ONE SIDE WITH HARDBOARD. SINK A PIECE OF WIRE MESH INTO THE MASS OF CANDY SO THAT IT STAYS IN PLACE."

"ALL YOU HAVE TO DO NOW IS TO PLACE YOUR CANDY ON THE INNER COVERS."

"I ADVISE YOU TO PROCEED IN SMALL QUANTITIES FOR YOUR FIRST TESTS. AS YOU PROGRESS, YOU'LL DECIDE FOR YOURSELF HOW MUCH YOU NEED. WHEN MANUFACTURING CANDY, MAKE SURE THAT HEATING THE INGREDIENTS IS SUFFICIENT AND FAST."

"OR! ... FOR THOSE WHO ARE STILL RELUCTANT TO MAKE CANDY, THEY CAN GET IT FROM A BEE STORE, BUT I SUGGEST A SIMPLE RECIPE (WHICH IS NEITHER CANDY NOR SYRUP) THAT BEES DON'T REJECT."

"SIR! CAN I HAVE A PIECE OF YOUR CANDY?"

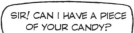

"FOR ONE HIVE: TAKE 35 OUNCES (1 KG) OF SUPERFINE SUGAR, WHICH YOU MELT IN JUST 1 CUP (0.25 LITER) OF WATER. HEAT. YOU OBTAIN A VERY SMOOTH SYRUP. MIX IN POWDERED SUGAR. WHEN THE PASTE REACHES A THICK CONSISTENCY, FILL YOUR MOLD AND SERVE THIS DISH TO YOUR BEES AS SOON AS THE NEED ARISES."

LATE FEEDING

PREPARING FOR WINTER

YOU TOOK ADVANTAGE OF A BEAUTIFUL AUTUMN DAY TO MAKE ONE LAST VISIT TO THE APIARY. IT WAS WITH PLEASURE THAT YOU NOTED THAT FOOD STORES WERE ABUNDANT AND THAT YOUR BEES WOULD SURVIVE THE WINTER WITHOUT FEAR OF STARVING. OTHERWISE, YOU JUST HAVE TO GIVE THEM SUGAR SYRUP.

NOW THAT THE PROBLEM OF FOOD STORES IS SOLVED, LET'S DEAL WITH THE OUTSIDE. YOU NEED TO CUT ANY BRUSH LIKELY TO OBSTRUCT THE ENTRANCE OF THE HIVE; PRUNE THE BRANCHES THAT COULD HIT THE WALLS IN HIGH WINDS.

DON'T FORGET TO INSTALL THE ENTRANCE REDUCERS OR MOUSE GUARDS, SINCE THE PRESENCE OF SMALL RODENTS INSIDE THE HIVE IS NOT DESIRABLE. YOU CAN MAKE THE WOODEN MODEL (A) YOURSELF; THE METAL ONE (B) IS AVAILABLE FROM BEE SUPPLY RETAILERS (THERE ARE DIFFERENT MODELS).

THE APPROACH OF WINTER REQUIRES VERY GREAT ATTENTION FROM THE BEEKEEPER. THE CARE GIVEN TO THE HIVES BEFORE THIS PERIOD WILL PRODUCE HEALTHY AND VIGOROUS COLONIES IN THE SPRING. WINTER MORTALITY CAN BE DUE TO A LACK OF FOOD.

THE AVERAGE CONSUMPTION FOR A COLONY HOUSED IN A DADANT-TYPE HIVE IS 33-44 POUNDS (15-20 KG) OF HONEY, AND IN A LANGSTROTH-TYPE HIVE 26-33 POUNDS (12-15 KG). IF YOU'RE WORRIED THAT WINTER WILL BE TOO LONG AND FOOD STORES WILL BE INSUFFICIENT DURING THE COLD SEASON, GIVE THEM BEE CANDY (THEY WILL BE GRATEFUL).

• PREPARING FOR WINTER •

LET'S OBSERVE THE BEHAVIOR OF A COLONY INSIDE THE HIVE DURING THE WINTER PERIOD. THE BEES ARE GROUPED TOGETHER. THOSE SURROUNDING THE CLUSTER BEGIN TO FLUTTER THEIR WINGS AS SOON AS THE TEMPERATURE REACHES 64°F (18°C). IN RESPONSE TO THIS RUSTLING, BEES CONSUME HONEY AND, AS A RESULT, RAISE THE TEMPERATURE IN THE HIVE. AS THESE CHANGES OCCUR, OTHERS REPLACE THE BEES THAT HAVE EATEN FROM THE CENTER. WHAT AN ORGANIZATION!

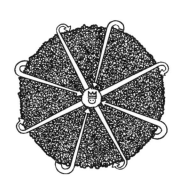

WHEN BEES FROM DIFFERENT COLONIES ARE UNABLE TO REACH THEIR HIVES, THEY GATHER TOGETHER TO ACCUMULATE THE HEAT NECESSARY FOR THEIR SURVIVAL. DURING WINTERING, AND ESPECIALLY IN VERY COLD WEATHER, AVOID MAKING NOISE AROUND THE HIVES. THE MERE FACT OF CLEARING THE ENTRANCES WOULD RISK DISRUPTING THE COLONY AND CAUSING EXCESSIVE HONEY CONSUMPTION. HOWEVER, THE RESERVES SHOULD NOT BE CONSUMED TOO QUICKLY IN CASE SPRING IS LATE. REMEMBER THAT THE AVERAGE WINTER DURATION FOR BEES IS SIX MONTHS, AND THE NECESSARY PROVISIONS ARE BETWEEN 50 AND 75 POUNDS (12 AND 15 KG) FOR A HIVE (DADANT OR LANGSTROTH MODEL).

WINTERING METHODS DIFFER FROM ONE COUNTRY TO ANOTHER, DEPENDING ON CLIMATIC CONDITIONS. IN THE UNITED STATES, SOME BEEKEEPERS PUT THEIR HIVES IN CELLARS. IN CANADA, THEY'RE GROUPED BY FOUR OR SIX AND LOCKED IN A KIND OF REMOVABLE BOX WITH OPENINGS CORRESPONDING TO THE ENTRANCES.

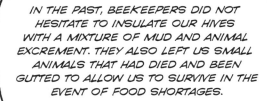

IN THE PAST, BEEKEEPERS DID NOT HESITATE TO INSULATE OUR HIVES WITH A MIXTURE OF MUD AND ANIMAL EXCREMENT. THEY ALSO LEFT US SMALL ANIMALS THAT HAD DIED AND BEEN GUTTED TO ALLOW US TO SURVIVE IN THE EVENT OF FOOD SHORTAGES.

NOWADAYS, SOME BEEKEEPERS LEAVE THE SUPERS ON THE HIVES, WHICH WE DON'T LIKE BECAUSE IT REDUCES THE INDOOR TEMPERATURE BY SEVERAL DEGREES. A SLIGHT DROP, YOU MIGHT SAY, BUT THESE FEW DEGREES ARE IMPORTANT TO US LITTLE BEES.

• WINTER HIVE VISIT •

"IF YOUR HIVES ARE IN A TEMPERATE REGION AND WINTER IS COMING TO AN END, ENJOY A NICE DAY OUT FOR A VISIT. GIVE A QUICK BANG AGAINST ONE OF THE WALLS OF THE HIVE. IF YOU HEAR WINGS RUSTLING VERY FAST, IT MEANS THAT EVERYTHING IS WELL."

"ALSO TAKE ADVANTAGE OF THIS VISIT TO CLEAR THE ENTRANCES OF THE HIVE SO THAT THE BEES AREN'T DISTURBED DURING THEIR FIRST TRIP OUTSIDE."

"IN THE COLDEST REGIONS, SNOW MAY HAVE BLOCKED THE ENTRANCE TO THE HIVE. AVOID CLEARING THE BOTTOM BOARD, SINCE YOUR ACT OF KINDNESS COULD DISRUPT THE CLUSTER. IF YOU FEEL IT'S NECESSARY TO CLEAR THE AREA, DO SO CAREFULLY AND JUDICIOUSLY."

"IF THE WINTER IS LONG AND YOU THINK YOUR BEES WILL RUN OUT OF FOOD, WHY DON'T YOU OFFER THEM A LOAF OF BEE CANDY THAT YOU PLACE ON THE INNER COVER, AVOIDING ANY SUDDEN MOVEMENTS."

"EXCUSE ME—CANDY CALLS."

A CLEAN BOTTOM BOARD

Since the appearance of the terrible Varroa mite, it's imperative to have clean bottom boards to be able to position the Varroa monitoring tray.

This monitoring tray consists of a frame on which a plastic or metal grid is stretched.

You may have noticed that bees always build wax constructions on the bottom boards and under the frames. In addition, ants bring humus and debris of all kinds. All this waste accumulates and ends up partially obstructing the entrance.

For those who have old hives, a real problem arises depending on whether they are equipped with an entrance reducer or a fixed bottom board. For fixed-board hives, I advise sawing the bottom. For those with entrance reducers, enlarge the openings. It's in your interest to equip your hives with reversible bottom boards.

Here are the tools needed to clean the bottom boards: a scraper, a wire brush, a blowtorch, and an empty super. For standard hives, one or two additional bottom boards are required.

SCRAPER

CHISEL

WIRE BRUSH

BLOWTORCH

• A CLEAN BOTTOM BOARD •

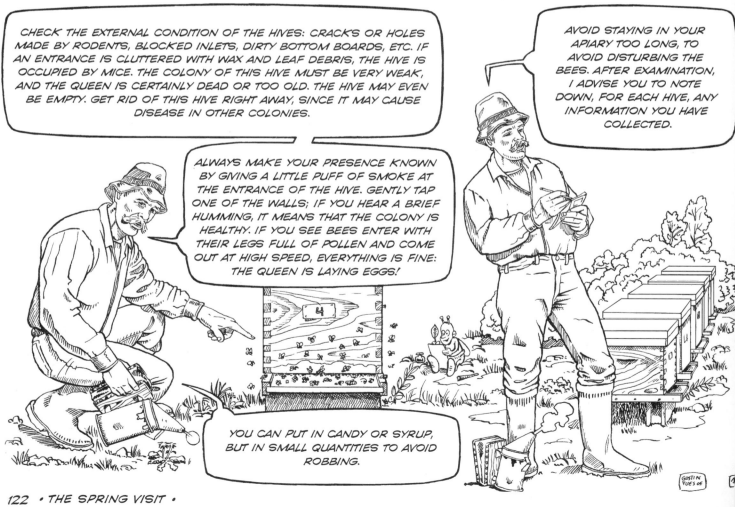

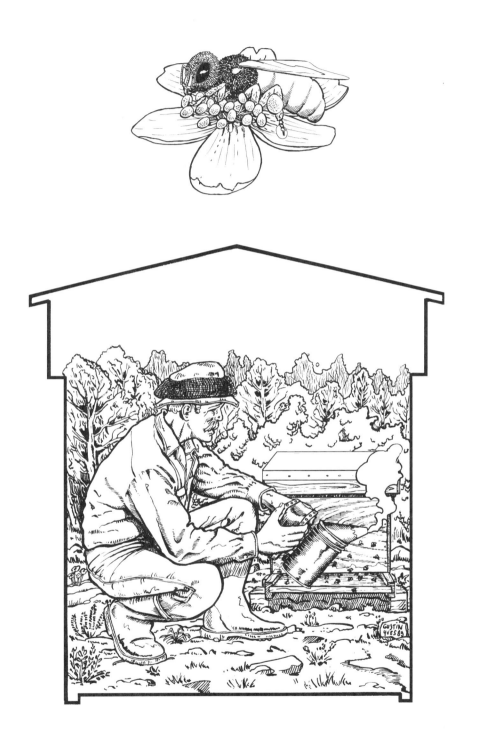

MAKING SYRUP

STIMULATE YOUR BEES WITH SYRUP.

Syrup, whether honey based or sugar based (and even a mixture of both), is eagerly welcomed by the bees. In autumn, the purpose of this food supply is to encourage wintering. The syrup will be thick enough to minimize ventilation efforts for water evaporation. The one given in spring, to stimulate egg laying, will be more liquid in order to provide the water necessary for brood rearing.

Recipe no. 1 (spring): Use 1 gallon water to 8.3 pounds of sugar.
Recipe no. 2 (fall): Use 1 gallon water to 16.6 pounds of sugar.

DIFFERENT COMMERCIAL FEEDERS

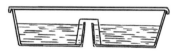

English feeder

Frame-cover feeder

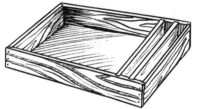

Frame-cover feeder made of wax-treated wood

Two plastic versions

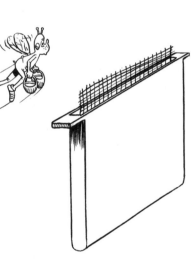

Plastic frame feeder with mesh ladder. You can make it with a frame, adding two sides made of hardboard and covering the whole thing in paraffin.

124 • MAKING SYRUP •

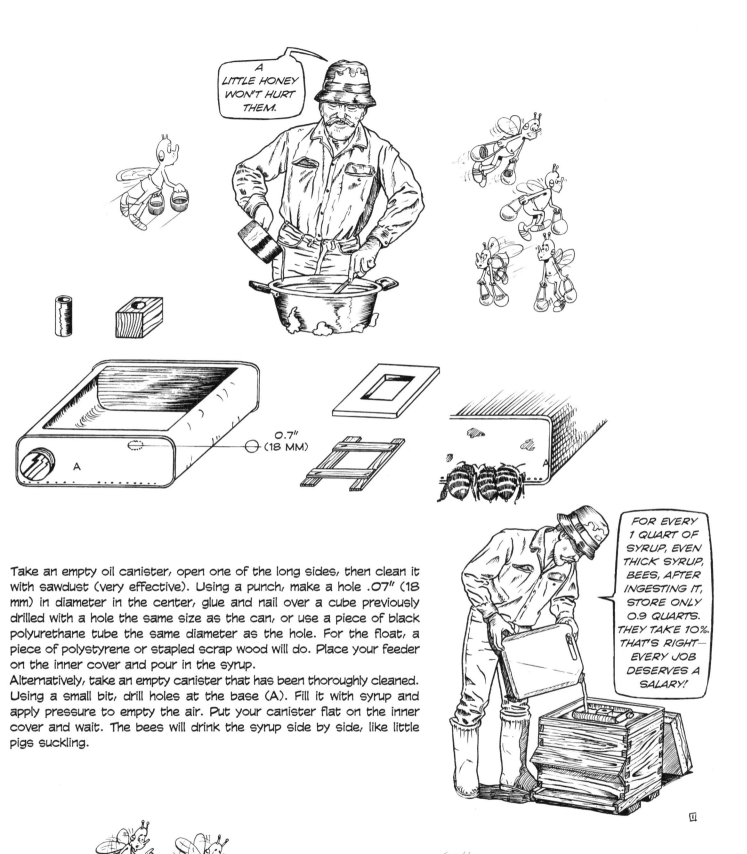

Take an empty oil canister, open one of the long sides, then clean it with sawdust (very effective). Using a punch, make a hole .07" (18 mm) in diameter in the center, glue and nail over a cube previously drilled with a hole the same size as the can, or use a piece of black polyurethane tube the same diameter as the hole. For the float, a piece of polystyrene or stapled scrap wood will do. Place your feeder on the inner cover and pour in the syrup.

Alternatively, take an empty canister that has been thoroughly cleaned. Using a small bit, drill holes at the base (A). Fill it with syrup and apply pressure to empty the air. Put your canister flat on the inner cover and wait. The bees will drink the syrup side by side, like little pigs suckling.

THE SWARMING PERIOD

IT'S SPRING! FLOWERING IS ABUNDANT, THE FIELDS ARE COVERED WITH A MULTITUDE OF FLOWERS, AND THE TREES ARE ADORNED WITH THEIR MOST BEAUTIFUL FINERY. A SMELL OF POLLEN AND NECTAR IS IN THE AIR AND CARESSES THE BOTTOM BOARDS.

BEES AREN'T OBLIVIOUS TO THIS CALL! WE CAN EXPECT THE FIRST SIGNS OF SWARMING FEVER. THE HOBBYIST OR PROFESSIONAL BEEKEEPER IS EAGERLY AWAITING THIS PERIOD, WHEN THEY WILL SEE THEIR BEES INCREASE IN NUMBERS OR TAKE OFF.

WHAT ARE THE MAIN CAUSES OF SWARMING?
1. HIGH POPULATION IN A SMALL SPACE WHERE THE QUEEN CAN NO LONGER LAY EGGS
2. LACK OF VENTILATION, WITH EXCESS HEAT
3. EXCESS NUMBER OF MALES
4. ABUNDANT NECTAR
5. I'VE OBSERVED THAT BEES, LIKE ANIMALS AND PLANTS, SWARM BY SURVIVAL INSTINCT FOLLOWING HARSH WINTERS, A PHENOMENON ALMOST NONEXISTENT AFTER A MILD WINTER.

"A SWARM OF BEES IN MAY IS WORTH A LOAD OF HAY. A SWARM IN JUNE IS WORTH A SILVER SPOON. A SWARM IN JULY ISN'T WORTH A FLY!"

LOOK AT THIS BEAUTIFUL SWARM. IT'S A PRIMARY SWARM COMPOSED OF THE OLD QUEEN WHO LEFT THE HIVE WITH BEES AND DRONES. IN GENERAL, IT LANDS A FEW METERS FROM THE HIVE. A SECONDARY SWARM, CONSISTING OF A VIRGIN QUEEN AND PART OF THE HIVE POPULATION—THUS WEAKENING THE HIVE—WILL BE MORE DIFFICULT TO COLLECT, WITH THE CAREFREE AND WILD YOUNG QUEEN LANDING ANYWHERE. MANY BEEKEEPERS DON'T CAPTURE SECONDARY SWARMS, BECAUSE THEY'RE NOT OF GREAT IMPORTANCE.

WE WON'T LEAVE EMPTY-HANDED! TAKE ALL THE POLLEN, HONEY, AND WAX WITH US.

TO HOBBYIST BEEKEEPERS WHO CATCH VERY SMALL SWARMS, I ADVISE YOU TO PUT A BROOD FRAME AND ANOTHER FRAME OF FOOD IN THE CENTER OF THEIR HIVE TO ENCOURAGE ITS GROWTH.

To avoid the loss of swarms, an ancestral method, but still practiced by beekeepers, consists of cutting off one of the wings of the fertilized queen to prevent it from flying away during swarming. This method is called clipping. Every beekeeper has the right to follow their swarm and retrieve it wherever it comes to rest. If the swarm has landed on private property, the beekeeper must ask the owners for permission to enter the property to recover it.

YOUR SWARMING PERIOD IS BETWEEN APRIL AND JUNE.

PLEASE LEAVE BETWEEN 11:00 A.M. AND 3:00 P.M.

OVER THE YEARS THAT I'VE BEEN OBSERVING SWARMS, I'VE NOTICED THEIR INTELLIGENCE. IT'S NOT THE QUEEN OR THE BEE THAT THINKS, BUT THE WHOLE COLONY. THE SWARM IS A SINGLE BEING WHO THINKS, LIVES, AND ACTS FOR THE GOOD OF THE COMMUNITY. FOR EXAMPLE, IF IT REALIZES THAT IT'S TOO SMALL, IT SIMULATES A FLIGHT AND RECOVERS BEES AND MALES FROM OTHER HIVES IN ITS WHIRLWIND. IT WILL REPEAT THIS OPERATION UNTIL IT CONSIDERS ITSELF BIG ENOUGH, AND IT WILL THEN DEFINITIVELY TAKE OFF IN SWARMING FLIGHT.

OUR BEEKEEPER FRIENDS HAVE ALWAYS USED MORE OR LESS EFFICIENT PROCESSES TO STOP US IN OUR SWARMING FLIGHT. SOME PEOPLE IMITATE THUNDER BY BANGING ON POTS OR CANS. OTHERS THROW DIRT OR SAND, AND EVEN WATER, TO MAKE US SETTLE DOWN.

IF HE THINKS THAT'S GONNA STOP US...

• THE SWARMING PERIOD •

CAPTURING A SWARM

• CAPTURING A SWARM •

• CAPTURING A SWARM • 133

INCREASING YOUR BEES

THIS VERY SIMPLE METHOD IS AIMED AT BEGINNERS OR WEEKEND BEEKEEPERS, WHO HAVE LITTLE TIME TO DEVOTE TO THEIR BEES, AND ALLOWS THEM TO MULTIPLY THEIR COLONIES WITHOUT TOO MANY RISKS. IT CONSISTS OF DIVIDING A COLONY (ON 10 FRAMES) IN TWO BY SEPARATING THE BROOD, HONEY, AND POLLEN FRAMES, AS WELL AS THE POPULATION. THE MOST FAVORABLE PERIOD TO CARRY OUT THIS OPERATION IS DURING SPRING ON A SUNNY DAY, WHEN THE DRONES ARE BORN AND WHEN THE COLONY POPULATION IS AVERAGE.

NOW IS THE TIME TO OPERATE. YOU HAVE PREPARED TWO EMPTY HIVES (FIVE TO SIX FRAMES).

MOVE THE MOTHER HIVE BACK AND REPLACE IT WITH THE TWO HIVES. IF YOU CAN'T MOVE YOUR MAIN HIVE, INSTALL THE NEW ONES IN FRONT OF IT. (DON'T FORGET TO SMOKE THE BEES FIRST.)

CAREFULLY REMOVE THE FIVE FRAMES FROM THE HIVE ONE BY ONE AND PLACE THEM CAREFULLY IN ONE OR THE OTHER OF THE HIVES. RESPECT THE ORDER IN WHICH YOU REMOVED THE FRAMES (WITH THE BEES); BE PREPARED TO IMPROVISE ACCORDING TO THE SITUATIONS (FOR EXAMPLE, WITH AN EXTRA BROOD FRAME).

AS SOON AS THE OPERATION IS FINISHED, COVER THE HIVE WITH A CANVAS OR WIRE FRAME LID. DO THE SAME FOR THE SECOND HIVE.

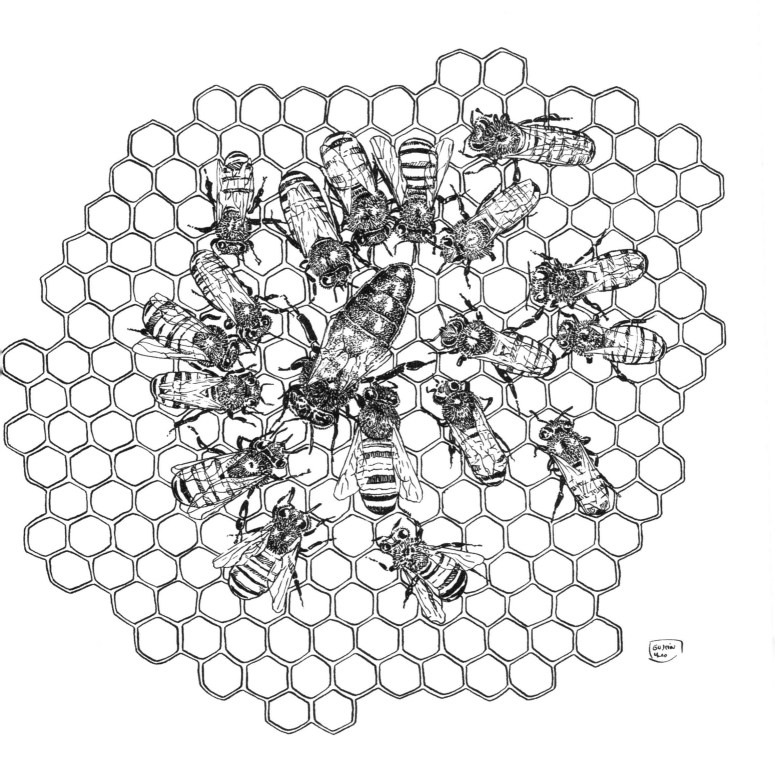

TRANSFERRING A HIVE

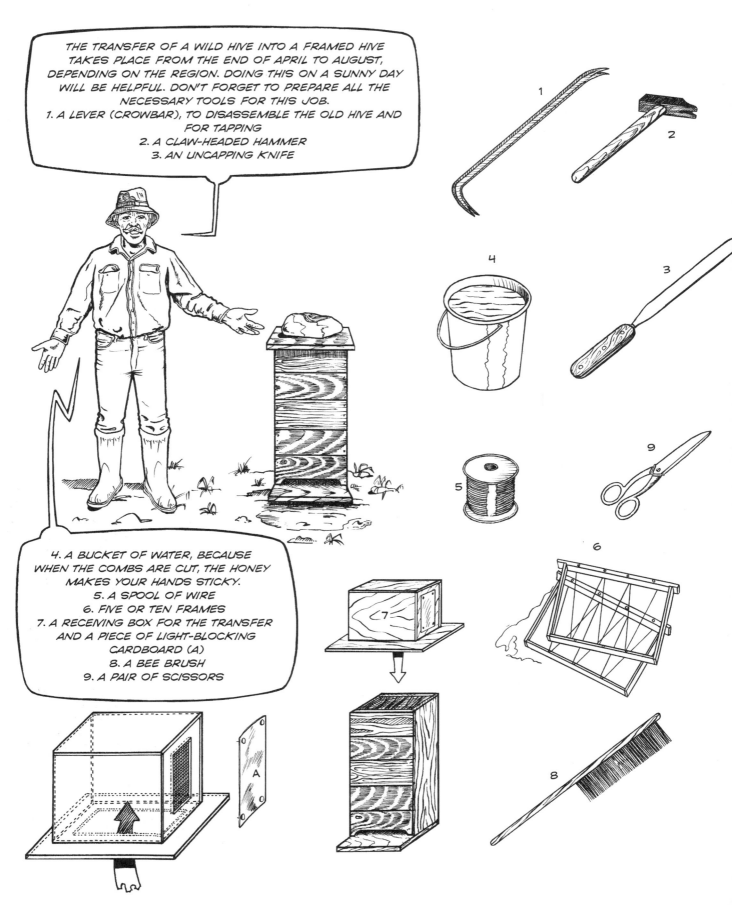

THE TRANSFER OF A WILD HIVE INTO A FRAMED HIVE TAKES PLACE FROM THE END OF APRIL TO AUGUST, DEPENDING ON THE REGION. DOING THIS ON A SUNNY DAY WILL BE HELPFUL. DON'T FORGET TO PREPARE ALL THE NECESSARY TOOLS FOR THIS JOB.
1. A LEVER (CROWBAR), TO DISASSEMBLE THE OLD HIVE AND FOR TAPPING
2. A CLAW-HEADED HAMMER
3. AN UNCAPPING KNIFE

4. A BUCKET OF WATER, BECAUSE WHEN THE COMBS ARE CUT, THE HONEY MAKES YOUR HANDS STICKY.
5. A SPOOL OF WIRE
6. FIVE OR TEN FRAMES
7. A RECEIVING BOX FOR THE TRANSFER AND A PIECE OF LIGHT-BLOCKING CARDBOARD (A)
8. A BEE BRUSH
9. A PAIR OF SCISSORS

142 • TRANSFERRING A HIVE •

• TRANSFERRING A HIVE •

BREEDING QUEENS

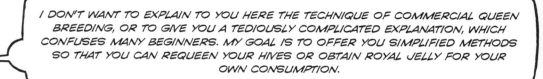

I DON'T WANT TO EXPLAIN TO YOU HERE THE TECHNIQUE OF COMMERCIAL QUEEN BREEDING, OR TO GIVE YOU A TEDIOUSLY COMPLICATED EXPLANATION, WHICH CONFUSES MANY BEGINNERS. MY GOAL IS TO OFFER YOU SIMPLIFIED METHODS SO THAT YOU CAN REQUEEN YOUR HIVES OR OBTAIN ROYAL JELLY FOR YOUR OWN CONSUMPTION.

HERE ARE FOUR POSSIBILITIES FOR NATURAL BREEDING:
1. WAIT FOR THE SWARMING PERIOD AND RECOVER THE QUEEN CELLS.
2. CAUSE SWARMING BY REDUCING THE LAYING AREA.
3. PLACE A QUEEN EXCLUDER TO DIVIDE THE HIVE.
4. REMOVE THE COLONY'S QUEEN AND "FORCE" BREEDING.

THE ESSENTIAL CONDITIONS FOR GOOD BREEDING ARE
— an abundance of food
— a good colony with a large number of young bees
— a favorable outside temperature (over 64°F/18°C)
— to start at the beginning of the season

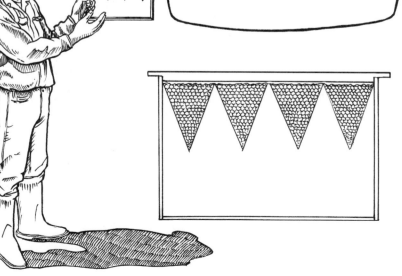

THE FIRST SIMPLE PROCEDURE I RECOMMEND CONSISTS OF PLACING A SHEET OF FOUNDATION WAX CUT INTO FOUR TRIANGLES (SEE DRAWINGS) IN AN EMPTY FRAME.

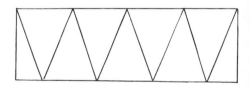

HERE YOU CAN SEE HOW TO CUT OUT THE SHEET OF FOUNDATION WAX WITHOUT LOSING ANY OF THE WAX, AND HOW TO PLACE IT.

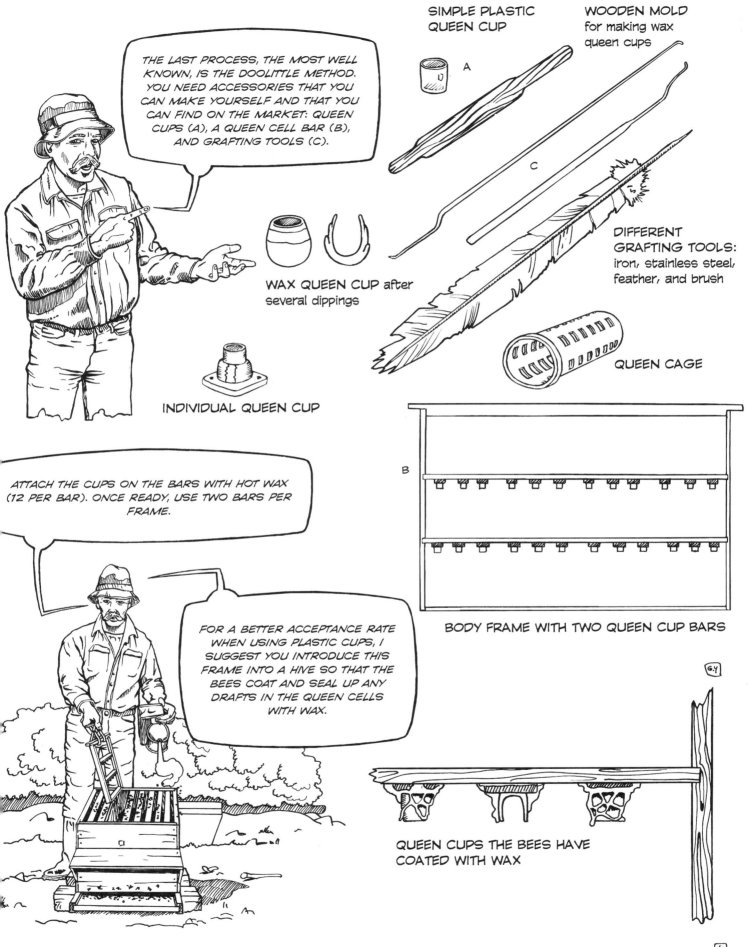

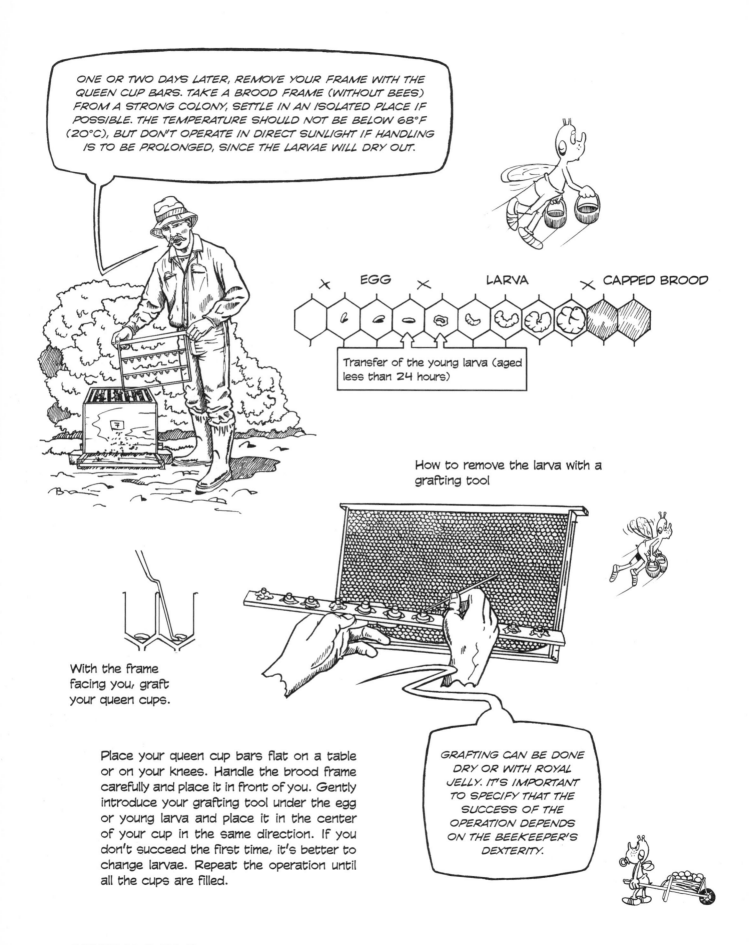

INTRODUCING QUEENS

YOUR COLONY DOESN'T HAVE A QUEEN; YOU WANT TO CHANGE YOUR QUEEN; YOU WANT TO TRY A NEW RACE: THOSE ARE THREE GOOD REASONS FOR INTRODUCING A QUEEN.

YOU HAVE TWO CHOICES FOR ACQUIRING A QUEEN:

1. YOU BREED YOUR OWN QUEEN BEES.

2. YOU WANT TO PURCHASE QUEENS. YOU'VE GOT PLENTY OF OPTIONS FOR FINDING GOOD QUALITY BREEDERS: ADS OR WORD-OF-MOUTH SUGGESTIONS FROM YOUR BEEKEEPING ASSOCIATION, LOCAL GROUPS, OR YOUR BEE INSPECTOR.

IF YOU HAVE CHOSEN TO BUY ONE OR MORE QUEENS, THEY WILL BE DELIVERED TO YOU IN THIS TYPE OF QUEEN CAGE (A) (BENTON CAGE IN GENERAL). IT'S DRILLED WITH THREE OR FOUR HOLES ACCORDING TO NEED, TO MAKE THE TRIP MORE COMFORTABLE, BUT ESPECIALLY FOR THE QUEEN'S SUBSISTENCE. IT'S FILLED WITH SIX TO TEN BEES FOR SMALL CAGES AND 10 TO 15 FOR LARGE ONES. THE FIRST HOLE IS FILLED WITH CANDY THAT IS USED FOR FOOD BUT ALSO FOR PLUGGING THE EXIT.

FOR THE INTRODUCTION OF YOUR QUEEN, I ADVISE YOU TO USE THIS SHIPPING CRATE. BE CAREFUL; SUCCESS IS NOT 100% GUARANTEED. WHATEVER THE METHOD YOU USE, FAILURE MAY BE DUE TO THE BEEKEEPER'S CLUMSINESS OR FEAR OF THE NEW QUEEN, EXCESSIVE USE OF SMOKE, OR INTRODUCTION DURING PERIODS OF FAMINE, BAD WEATHER, ETC. IF YOU MANIPULATE THE QUEEN BETWEEN YOUR FINGERS, YOU MAY SEE HER BALLED BY THE OTHER BEES.

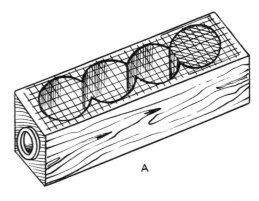

A

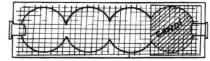

WHY AND HOW IS A QUEEN BALLED?
First of all, the longer your colony has been without a queen, the greater the risk of failure, since they are the most defensive old bees. On the other hand, the younger the bees are and the younger the queen is, the more likely she will be accepted. It's enough for the introduced queen to be afraid or to emit an unusual smell for an old bee to attack her and trigger a murderous "madness" on the part of the old bees. The queen is surrounded by agitated bees who try to suffocate her; this merciless fury materializes in a ball; hence the expression of queen balling. If you witness this scene, smoke the ball or throw it in the water to disperse the bees and recover the queen, if it's not too late!

BEFORE INTRODUCTION, CHECK THAT YOUR HIVE IS QUEENLESS AND THAT IT HAS NO QUEEN CELLS OR LAYING WORKERS. OTHERWISE, YOUR NEW QUEEN WILL NOT BE ACCEPTED.

IF YOU WANT ONLY TO CHANGE QUEENS, THERE IS A SIMPLE AND EFFECTIVE METHOD: PLACE THE OLD QUEEN IN A QUEEN CAGE AND PUT HER BACK IN THE HIVE UNTIL THE NEXT DAY. AFTER THIS TIME, DESTROY THE OLD QUEEN AND REPLACE IT WITH THE NEW ONE, WHICH WILL TAKE ON THE SMELL LEFT IN THE CAGE BY THE OLD QUEEN.

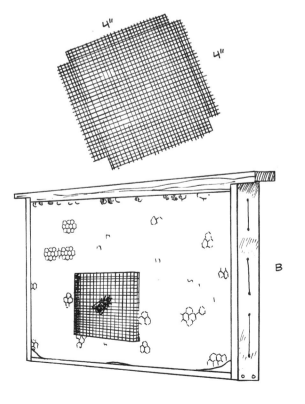

No matter what method you use, it's advisable to remove accompanying nurse bees before introduction.

THIS IS A FIRST ADVISORY PROCEDURE FOR BEEKEEPERS BREEDING THEIR OWN QUEENS. TAKE YOUR CAGE AND SOAK IT IN WATER AT ROOM TEMPERATURE TO PREVENT THE QUEEN FROM FLYING AWAY. PLACE THE QUEEN ON A RECENT BROOD FRAME. TRAP IT WITH A 4" (10 CM) SQUARE OF MESH (B). PUT YOUR FRAME BACK IN THE QUEENLESS HIVE. TWO TO THREE DAYS LATER, RELEASE YOUR QUEEN, WHO HAS BEEN FED BY YOUNG BEES THROUGH THE WIRE MESH AND THUS ADOPTED BY THE COLONY.

• INTRODUCING QUEENS • 153

JOINING

FOR BEGINNERS, IT'S SOMETIMES DIFFICULT TO PART WITH A WEAK COLONY AND SEE A DECREASE IN BEES. BUT A WEAK COLONY REQUIRES AS MUCH WORK AS A STRONG ONE AND GIVES NOTHING IN RETURN. THAT'S WHY IT'S BETTER TO JOIN TWO WEAK COLONIES IN THE FALL, BECAUSE BY THAT TIME IT'S TOO LATE TO CHANGE THE QUEEN.

THERE ARE DIFFERENT METHODS FOR BRINGING TWO COLONIES TOGETHER. I'M GOING TO SHOW YOU THE SIMPLEST AND FASTEST, BECAUSE COMPLICATED METHODS DON'T ALWAYS GIVE THE EXPECTED RESULTS.

OPERATE IN THE LATE AFTERNOON. SMOKE A LITTLE MORE THAN USUAL IN ORDER TO WORK WITH CALM BEES: THE SMOKE WILL ENCOURAGE THEM TO STUFF THEMSELVES WITH HONEY, MAKING THEM LESS DEFENSIVE.

FOR THE BEST RESULTS, IT'S BETTER TO BRING TOGETHER TWO NEIGHBORING COLONIES. IF THEY'RE FAR FROM EACH OTHER, BUT WITHIN THE SAME APIARY, MOVE ONE OF THE TWO HIVES GRADUALLY, EACH EVENING. IF ONE OF THE COLONIES IS IN ANOTHER APIARY (MORE THAN 2 MILES / 3 KM AWAY), BRING IT IN ONE MOVE.

THE OPERATION IS VERY SIMPLE: PLACE A SHEET OF NEWSPAPER ON ONE OF THE TWO HIVES, PLACE THE BODY OF THE OTHER ON TOP (WITHOUT THE BOTTOM BOARD, OF COURSE), AND MAKE A FEW HOLES IN THE NEWSPAPER SHEET. THE BEES WILL HAVE ALL NIGHT TO MIX THEIR SCENTS AND GET TO KNOW EACH OTHER BETTER.

A FEW DAYS LATER, REMOVE THE UNNECESSARY FRAMES FROM BOTH BODIES AND GROUP THE BROOD FRAMES IN THE LOWER BODY.

DO NOT LEAVE THE EMPTY HIVE IN THE APIARY.

WHATEVER METHOD IS CHOSEN, YOU'LL PUT ALL THE CHANCES ON YOUR SIDE BY REMOVING THE WEAKEST QUEEN, AND YOU'LL HIRE ME, THE NEW QUEEN (SEE THE CHAPTER ON "INTRODUCING QUEENS").

HONEY

NOW LET'S TALK ABOUT THE MOST IMPORTANT HIVE PRODUCT FOR THE BEEKEEPER AND THEIR BEES: HONEY. WE WILL REVIEW THE FOLLOWING TOPICS: THE ORIGIN, MANUFACTURE, COMPOSITION, AND DIFFERENT TYPES OF HONEY.

A LITTLE HISTORY TO START WITH

Without bees, there's no honey! Primitive humans (Paleolithic era) did not hesitate to steal honey from bees to feed themselves (a little like bears), as some rock paintings testify. Documents dating back to 1600 BCE revealed that children were fed and treated for minor ills with honey. Did you know that the Egyptians were buried not only with their wealth but also with honey? Beekeeping developed in ancient Greece. Then, with the arrival of sugar cane in Europe, honey lost its prestige. Under Napoleon, sugar beet flourished with the continental blockade. Because of this, honey became increasingly rare on the table.

LET ME TAKE A MOMENT TO ANALYZE THE DIFFERENCES AMONG HONEY, CANE SUGAR, AND BEET SUGAR.

Cane and beet sugars are nothing more or less than sucrose with added sucrose (i.e., they can't be assimilated directly by the body), while honey is made up of fructose, glucose, vitamins, minerals, trace elements, and diastases. Sugar, a dead food, is a source of germs, while honey, a living food, cannot be a carrier of germs because it contains antimicrobial substances.

WHERE DOES HONEY COME FROM? Let's start at the beginning. The earth, the plant, the insect, plus two elements (sun and water) constitute a small natural factory.

LET'S SEE HOW THE CENTRAL NUCLEUS OF THIS "FACTORY" WORKS: THE PLANT. IT DRAWS ITS ENERGY FROM THE SUN'S RAYS THROUGH CHLOROPLASTS THAT CONTAIN CHLOROPHYLL, FOUND IN THE CELLS OF ITS LEAVES. THIS PROCESS IS CALLED PHOTOSYNTHESIS. BY MEANS OF THE PUMP (THE ROOT), THE PLANT DRAWS MINERAL SALTS, NITRATES, AND WATER FROM THE SOIL AND USES ITS STEM AS A FOOD CARRIER. TO PERPETUATE THE SPECIES, IT USES ONE MEANS AMONG OTHERS: THE FLOWER. THE LOVELY COLORS OF FLOWERS ARE THERE TO ATTRACT BEES, WHICH TAKE THE SMALL DROP OF NECTAR FOUND AT THE BOTTOM OF THE FLOWER'S "CUP."

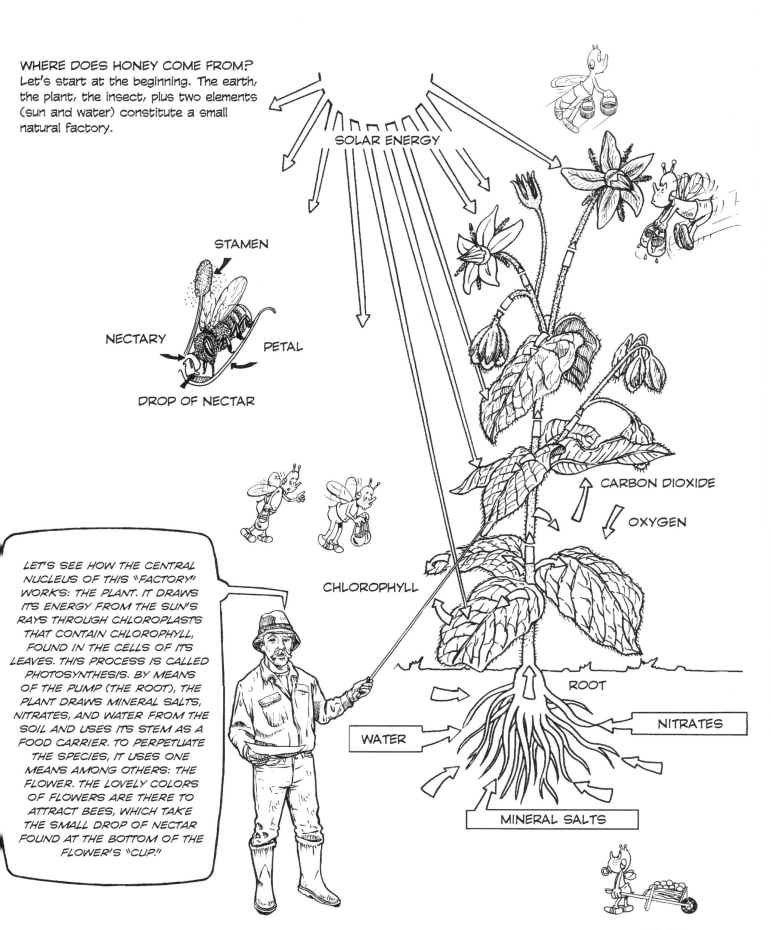

• HONEY •

WATCH US AT WORK, MY FRIEND!

A bee spots a pretty flower.

It settles on it and sucks up a drop of nectar, which at this stage contains about 50% to 80% water. When its crop is full, the bee returns to its hive. During the journey, the nectar loses half its weight in water (by evaporation). Thus concentrated, it receives juice from a frontal gland, and by a back-and-forth movement between pharynx and crop, it's mixed with diastase and transformed into honey.

When the bee arrives at the hive, it disgorges the honey in a cell by muscle contractions, then goes off again to forage until the cell is completely filled. When the honey has a reduced moisture content (15%—20%), the cell is covered with a thin layer of wax.

LOCUST	CANOLA	SAINFOIN	HORSE CHESTNUT	CLOVER AND ALFALFA
✿ May—June	✿ May—June	✿ June	✿ June—July	✿ June—July
☐ fluid	☐ compact	☐ creamy	☐ thick	☐ creamy
△ light	△ light	△ white	△ dark	△ white

✿ Flowering season   ☐ Honey consistency   △ Honey color

HERE'S A SUMMARY OF THE COMPOSITION OF HONEY AND ITS MAIN PROPERTIES. THIS DATA IS ONLY APPROXIMATE AND VARIES DEPENDING ON THE REGIONS AND FLOWERS VISITED.

EVERYTHING OLD IS NEW AGAIN ... FOR EXAMPLE, PACKAGING HONEY IN SQUEEZE TUBES. THE IDEA ISN'T NEW; BEFORE 1950, A CERTAIN JOHN F. HAWKINS, FROM CHESTER, PENNSYLVANIA, WAS ALREADY MARKETING HONEY IN TUBES.

HONEY COMPOSITION
Water 18% – Carbohydrates 78% (laevulose glucose and 1%–2% sucrose) – Mineral salts – Trace elements – Vitamins B1, B2, B3 – Digestive enzymes

HONEY PROPERTIES
Energetic – Revitalizing – Toning – Antibiotic – Recalcifying. Each honey has specific properties depending on the flowers foraged.

PACKAGING HAS CHANGED A LOT SINCE THE BEGINNING OF THE 20TH CENTURY, WHEN SHOPKEEPERS CARRIED COMB HONEY. THE QUANTITY THE CUSTOMER WANTED WAS CUT OFF WITH CHEESE WIRE AND WRAPPED IN GREASEPROOF PAPER.

NOWADAYS, PACKAGING CONTINUES TO EVOLVE, SOMETIMES CHANGING FOR ECOLOGICAL REASONS. AT BEE SUPPLY RETAILERS YOU'LL FIND A WIDE VARIETY OF OPTIONS TO CONSIDER, FROM GLASS TO PLASTIC TO PAPER CONTAINERS.

I DON'T KNOW IF YOU'RE LIKE ME, BUT WHEN I HOLD A JAR OF HONEY IN MY HANDS, I CAN'T RESIST DIGGING IN.

IF YOU WANT TO KEEP YOUR HONEY FOR A LONG TIME, KEEP IT AT A CONSTANT TEMPERATURE OF 57°F (14°C). IF YOU PREFER A MORE LIQUID HONEY, MELT IT IN A DOUBLE BOILER, BUT BE CAREFUL: THE WATER SHOULD NOT BOIL! (YOU SHOULD BE ABLE TO PUT A FINGER IN IT WITHOUT GETTING BURNED.)

MEAD

POLLEN

POLLEN IS FOUND ON THE STIGMA AND ALLOWS THE FLOWERS TO BE FERTILIZED. THE PROFILE OF A POLLEN GRAIN IS DIFFERENT DEPENDING ON THE PARTICULAR FLOWER IT COMES FROM (IT CAN BE SPHERICAL, OVOID, CUBIC, ELONGATED, ETC.). IT CONTAINS A LARGE NUMBER OF VITAMINS (ALL B) AND A LARGE NUMBER OF ENZYMES, CARBOHYDRATES, LOW FAT, ALBUMINOUS SUBSTANCES, LIME AND MAGNESIUM PHOSPHATES, ANIMAL GELATIN, AND MALIC ACID.

BEEKEEPING IS NOT JUST ABOUT HARVESTING HONEY. IT'S AN ESSENTIAL FACTOR IN POLLINATION. FORAGERS ALLOW POLLEN TO BE TRANSMITTED FROM ONE FLOWER TO ANOTHER, THUS PROMOTING FERTILIZATION. WHILE POLLEN IS USEFUL AND EVEN ESSENTIAL TO BEES, THAT'S NOT THE CASE FOR SOME HUMAN BEINGS WHO ARE ALLERGIC TO IT AND SUFFER FROM THE CONDITION OFTEN CALLED HAY FEVER.

SOME PLANT VARIETIES WOULD NOT BEAR FRUIT WITHOUT THE BENEFICIAL ACTION OF BEES. PLANTS FALL INTO TWO MAIN CATEGORIES: CRYPTOGAMS AND PHANEROGAMS. LET'S LOOK AT THE SECOND CATEGORY, BECAUSE THE FIRST ONE HAS HIDDEN ORGANS. LET'S TAKE THE EXAMPLE OF A NON-AUTO-FERTILIZING CHERRY BLOSSOM THAT CONSISTS OF MALE AND FEMALE ORGANS. THE BEE THAT VISITS THIS FLOWER DROPS THE POLLEN GRAINS ONTO THE STAMENS, THUS ALLOWING FERTILIZATION. SOME CHERRY (OR OTHER FRUIT) SPECIES AREN'T SELF-FERTILIZING, AND THEIR FRUIT PRODUCTION DEPENDS ENTIRELY ON THE PRESENCE OF POLLINATING INSECTS (BEES, AMONG OTHERS).

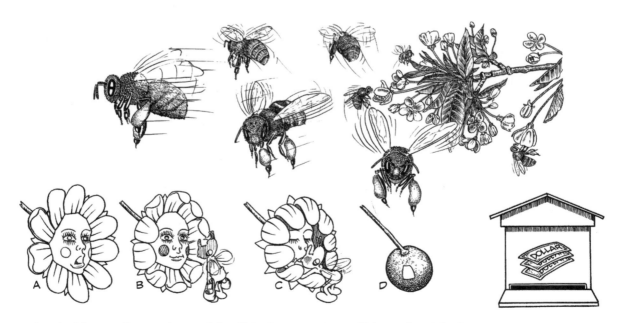

One morning in May, a flower dressed itself in its most beautiful petals with the intention of pleasing a forager (A). A bee, attracted by the beauty of the flower, came to visit it and set out again loaded with pollen (B). Theft, do you think? No, simply an exchange, because in return, the bee, which carried the pollen necessary for brood rearing back to the hive, made it possible for the flower (C) to produce a seed or fruit (D).

In many countries, beekeepers are paid to bring their hives into orchards and fields to fertilize the flowers.

176 • POLLEN •

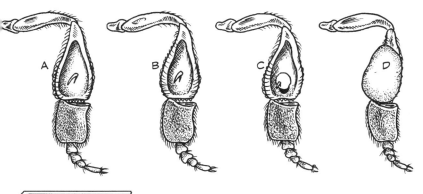

Let's analyze the hind leg of the bee that allows it to carry pollen. The upper part of this organ is carved in the shape of a basket (A); it has long hairs that evoke claws (B). In the center, we can see a hook (C) around which the bee will pile up the pollen (D). The spur (E) of the middle leg is used to remove the balls of pollen.

When the forager feels that her pollen load is sufficient, she goes to the hive and deposits her supply in a cell (1). She then goes back to work, leaving it to a young bee to store the pollen; she will pack the balls of pollen at the bottom of the cell until the cell is full (2). Finally, the young worker covers everything with a thin layer of honey to avoid fermentation.

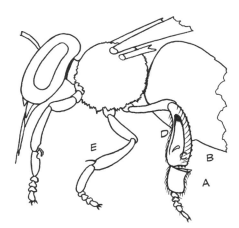

Let's look at a bee's legs when harvesting pollen. Using its front legs, the bee collects the pollen from the stamen and then, using the spur (C), comb (A), and claws, places it in the baskets (D) on the hind legs. To form the balls, the pollen is mixed with nectar before being placed in the baskets. These operations are carried out very quickly and in full flight.

POLLEN IS BEES' ONLY SOURCE OF PROTEIN.

The average consumption for an average colony is 55 pounds (25 kg) of pollen per year, and 77 pounds (35 kg) for a strong colony.

• POLLEN • 177

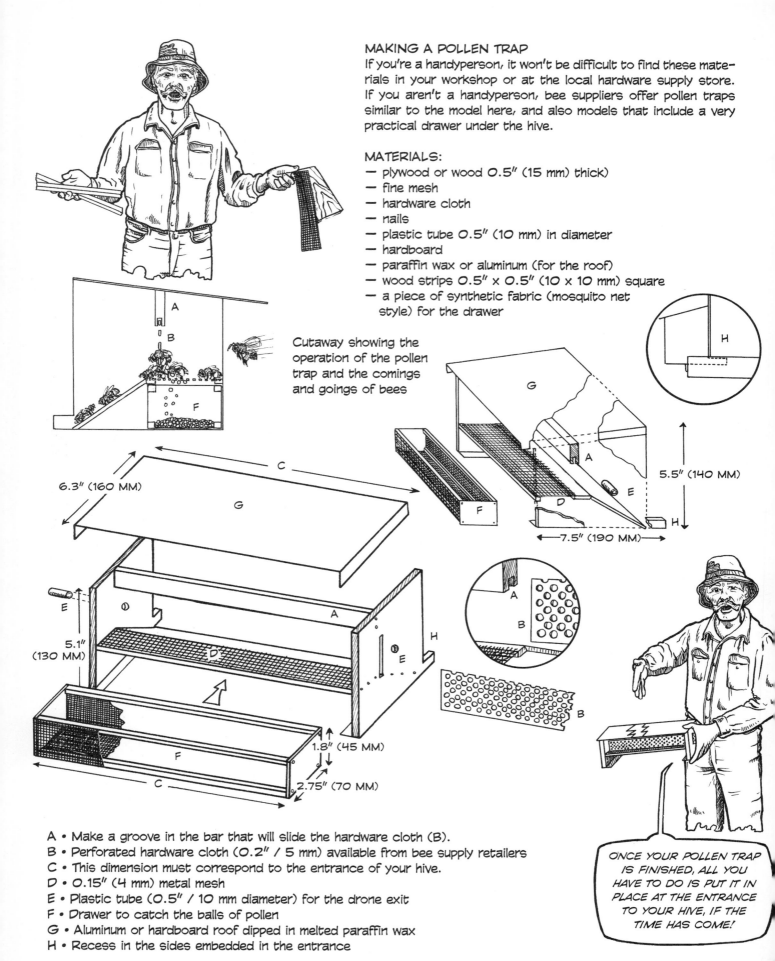

MAKING A POLLEN TRAP

If you're a handyperson, it won't be difficult to find these materials in your workshop or at the local hardware supply store. If you aren't a handyperson, bee suppliers offer pollen traps similar to the model here, and also models that include a very practical drawer under the hive.

MATERIALS:
- plywood or wood 0.5" (15 mm) thick)
- fine mesh
- hardware cloth
- nails
- plastic tube 0.5" (10 mm) in diameter
- hardboard
- paraffin wax or aluminum (for the roof)
- wood strips 0.5" x 0.5" (10 x 10 mm) square
- a piece of synthetic fabric (mosquito net style) for the drawer

Cutaway showing the operation of the pollen trap and the comings and goings of bees

A • Make a groove in the bar that will slide the hardware cloth (B).
B • Perforated hardware cloth (0.2" / 5 mm) available from bee supply retailers
C • This dimension must correspond to the entrance of your hive.
D • 0.15" (4 mm) metal mesh
E • Plastic tube (0.5" / 10 mm diameter) for the drone exit
F • Drawer to catch the balls of pollen
G • Aluminum or hardboard roof dipped in melted paraffin wax
H • Recess in the sides embedded in the entrance

ONCE YOUR POLLEN TRAP IS FINISHED, ALL YOU HAVE TO DO IS PUT IT IN PLACE AT THE ENTRANCE TO YOUR HIVE, IF THE TIME HAS COME!

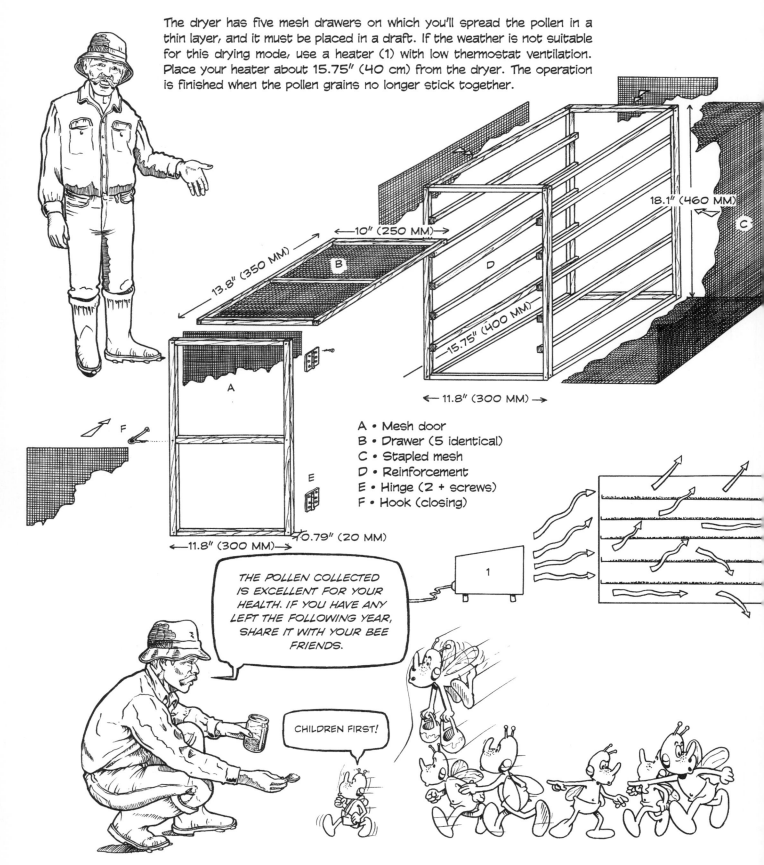

MAKING A POLLEN DRYER

Material:
- Wood strips 0.8" x 0.8" (20 x 20 mm) for the frame and door, 0.5" x 0.5" (10 x 10 mm) for the drawers
- Netting, such as mosquito net or food-safe mesh

The dryer has five mesh drawers on which you'll spread the pollen in a thin layer, and it must be placed in a draft. If the weather is not suitable for this drying mode, use a heater (1) with low thermostat ventilation. Place your heater about 15.75" (40 cm) from the dryer. The operation is finished when the pollen grains no longer stick together.

A • Mesh door
B • Drawer (5 identical)
C • Stapled mesh
D • Reinforcement
E • Hinge (2 + screws)
F • Hook (closing)

THE POLLEN COLLECTED IS EXCELLENT FOR YOUR HEALTH. IF YOU HAVE ANY LEFT THE FOLLOWING YEAR, SHARE IT WITH YOUR BEE FRIENDS.

CHILDREN FIRST!

ROYAL JELLY

THIS SMALL BOTTLE I'M HOLDING CONTAINS 3 GRAMS OF ROYAL JELLY, A GELATINOUS AND WHITISH MATERIAL WITH A SLIGHTLY ACIDIC TASTE. ROYAL JELLY, OR BEE MILK, IS THE QUEEN'S EXCLUSIVE FOOD THROUGHOUT HER LIFE, AND THE MAIN FOOD FOR ALL LARVAE UNTIL THE THIRD DAY OF THEIR DEVELOPMENT (INCREASED 1,000 TIMES IN THREE DAYS FOR WORKER LARVAE AND 2,000 TIMES IN FIVE DAYS FOR QUEEN LARVAE).

THREE CIRCUMSTANCES ALLOW THE BEEKEEPER TO HARVEST ROYAL JELLY:
1. DEATH OF THE QUEEN; 2. SWARMING (NATURAL OR ARTIFICIAL);
3. QUEEN REMOVED BY THE BEEKEEPER.
WHEN ONE OF THESE THREE CASES ARISES, THE WAX WORKERS BUILD QUEEN CELLS FROM CELLS CONTAINING BROOD LESS THAN 48 HOURS OLD. THE NURSES THEN FEED THE LARVAE ROYAL JELLY UNTIL THE QUEEN CELL IS SEALED.

WHY DO BEES BUILD A QUEEN CELL?

The queen secretes a queen substance that is released from the entire surface of her body. Young workers lick the queen's body to soak up the smell, which they then pass on to the entire colony. As long as the pheromone, the main component of the queen substance, is diffused in the hive, the workers don't build queen cells. On the other hand, if the queen substance decreases or disappears, then the process of building the queen cells is triggered to replace the deficient or missing queen. It should be noted that if the queen's laying slows down, it does not in any way influence the building of the queen cells. The queen substance of queens is not identical to that of virgin queens, and only bees know the difference.

IT'S THE BEES AGED FROM 5 TO 14 DAYS WHO PRODUCE THIS ROYAL JELLY, THANKS TO THEIR FRONTAL GLANDS, OR PHARYNGEAL, WHICH DEVELOP AT THIS PRECISE STAGE OF THEIR EXISTENCE ...

... AND WHICH LATER ATROPHIES. IT'S ONLY DURING THIS PERIOD THAT THE YOUNG BEE FUNCTIONS AS A NURSE.

1. PHARYNGEAL GLANDS

The pharyngeal glands are located on the right and left of the head; they form a small string of 500 lymph nodes.

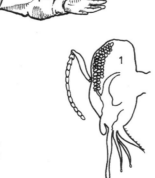

A • Pharyngeal glands developed on a nurse bee
B • Atrophied glands of a bee that is no longer a nurse

THE EASIEST WAY FOR THE HOBBYIST BEEKEEPER TO OBTAIN ROYAL JELLY IS TO WAIT FOR THE SWARMING PERIOD, WHICH IS, DEPENDING ON THE YEAR, FROM APRIL TO JUNE. AT THAT TIME, YOU'LL FIND QUEEN CELLS LIKE THESE IN YOUR HIVES. AFTER VISITING YOUR COLONIES AND JUDGING THEIR CONDITION, YOU'LL DECIDE ON THE TOTAL OR PARTIAL REMOVAL OF THE QUEEN CELLS. THEY'RE THE ONES THAT PROVIDE ROYAL JELLY.

A BEEKEEPER WHO WANTS TO HARVEST ROYAL JELLY IN LARGER QUANTITIES USES ONE OF TWO METHODS: EITHER BY DESTROYING THE QUEEN OR BY TEMPORARILY ISOLATING HER. BE CAREFUL, BECAUSE AS SOON AS THE BEES NOTICE THE DISAPPEARANCE OF THEIR QUEEN, THE WAX MAKERS WILL HURRY TO BUILD QUEEN CELLS. TO OVERCOME THIS DISADVANTAGE, REMOVE THE YOUNG BROOD FRAMES AND INSERT FRAMES WITH QUEEN CUPS (WAX, WOOD, OR PLASTIC) IN WHICH YOU HAVE GRAFTED LARVAE LESS THAN 48 HOURS OLD (GRAFTING CAN BE DONE "DRY" OR WITH A DROP OF ROYAL JELLY). THREE DAYS LATER, REMOVE THE FRAME TO HARVEST THE ROYAL JELLY CONTAINED IN THE CELLS. THIS IS CARRIED OUT WITH A SPATULA OR BY SUCTION.

Place more or fewer cups, depending on your needs, knowing that a queen cell contains about 200 mg of jelly. Pack your harvest without delay in aseptic and hermetically sealed vials that you place away from light and at a temperature of 32°F–41°F (0°C–5°C).

A • Queen cell with wax makers (1) and nurse bees (2)
B • Plastic cup with grafted larva

LINE UP HERE FOR SOME ROYAL JELLY.

PROPERTIES

Stimulating
Rejuvenating
Toning
Rebalancing
Revitalizing
Euphoria inducing

COMPOSITION

70% water
30% dry matter, of which the main elements are proteins; carbohydrates; lipids; trace elements; amino acids; vitamins B1, B2, B2, B3, B5, B6, B7, B8, B9, and B12, as well as A, C, D, and E; plus substances that are still unknown.

• ROYAL JELLY • 183

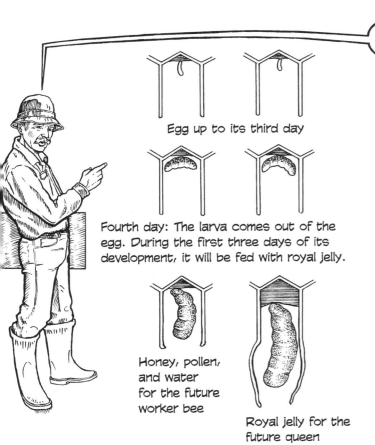

Egg up to its third day

Fourth day: The larva comes out of the egg. During the first three days of its development, it will be fed with royal jelly.

Honey, pollen, and water for the future worker bee

Royal jelly for the future queen

TO PROVE THE EXCEPTIONAL VALUE OF ROYAL JELLY, LET'S TAKE THESE TWO CELLS CONTAINING TWO IDENTICAL LARVAE. FED WITH ROYAL JELLY, THEY SEE THEIR INITIAL WEIGHT INCREASE BY 1,000 TIMES DURING THE FIRST THREE DAYS OF THEIR DEVELOPMENT. AT THIS STAGE, THE LARVA (A), A FUTURE WORKER, IS FED A MIXTURE OF WATER, HONEY, AND POLLEN. FOR SPECIAL REASONS, LARVA (B) IS CHOSEN TO BECOME A QUEEN; SHE IS PAMPERED BY THE NURSES, WHO FILL ITS CELL WITH ROYAL JELLY, WHOSE LAYER CAN REACH 1 CM. THE ABSORPTION OF THIS JELLY ALLOWS A CONSIDERABLE DEVELOPMENT OF THE FUTURE QUEEN'S REPRODUCTIVE SYSTEM, AND SHE WILL BE ABLE TO LAY UP TO 2,000 EGGS PER DAY.

THE BENEFITS OF ROYAL JELLY DON'T END THERE. THIS PRECIOUS SUBSTANCE IS A GUARANTEE OF LONGEVITY: A WORKER LIVES FOUR TO SIX WEEKS IN SUMMER AND FOUR TO SIX MONTHS IN WINTER, WHILE A QUEEN LIVES FOUR TO SIX YEARS ON AVERAGE.

LET'S LOOK BACK AT ONE OF THE IMPORTANT COMPONENTS OF ROYAL JELLY: VITAMIN B5, OR PANTOTHENIC ACID. THIS ACID IS ESSENTIAL FOR THE ASSIMILATION OF DIGESTIBLE PRODUCTS BY OUR BODY. A VITAMIN B5 DEFICIENCY CAN HAVE IMPORTANT CONSEQUENCES ON OUR BODY: INTELLECTUAL AND PHYSICAL FATIGUE, DIGESTIVE AND INTESTINAL DISORDERS, HAIR LOSS, SKIN DISEASES, ANEMIA, ETC. THE LEVEL OF PANTOTHENIC ACID IN ROYAL JELLY IS THE HIGHEST FOUND IN A NATURAL PRODUCT.

LABORATORIES MARKET ROYAL JELLY FOR DIFFERENT USES. THEY'RE SOLD IN FORMATS SUCH AS CAPSULES AND TABLETS, AND IN ALL SORTS OF BEAUTY PRODUCTS.

I WANT TO BE A QUEEN TOO.

SOME TIPS FOR STORING ROYAL JELLY
Ideally, it should be consumed as soon as it's harvested. Pure, it must be stored in the refrigerator. Between 32°F and 41°F (0°C—5°C), it can be stored for several months. Mixed with honey (3 g per 125 g of honey), it should be stored in the refrigerator or at a temperature not exceeding 47°F (14°C) and protected from light. It can be kept for a little over a year, but over time its active ingredients decrease.

WAX

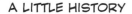

"THE WAX CAN BECOME A BLOCK OR A FOUNDATION WAX SHEET."

A LITTLE HISTORY

In ancient Greece, wax was a material used to represent the deities in sculpture, and wax tablets were used as writing materials.

According to legend, Icarus wanted to escape from the labyrinth, and he made wings by gluing feathers together with wax. Unfortunately, he flew so high that the sun melted the wax, and, without his wings, Icarus fell and drowned in the Aegean Sea.

Much later, physicists and philosophers believed that wax came from plants. Error! After long observation, Hunter noticed that bees made wax. Duchet and Hubert made identical observations.

The 11-day-old bees are responsible for producing the wax thanks to their wax glands (A). This function keeps them very busy and lasts for only about 10 days.

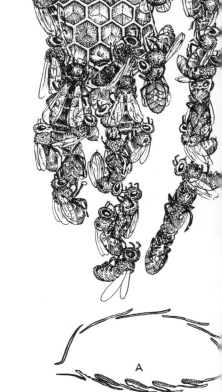

"YOU HAVE STORED YOUR CAPPINGS EITHER IN BAGS OR CANS. NOW IT'S UP TO YOU TO MAKE GOOD USE OF THEM."

It's thanks to the help of its legs that the bee recovers the wax scales emanating from its abdomen. It manipulates them and then gives them to another worker. Gradually, the cells are built.

A "WAX FACTORY" BRINGS TO MIND A SOPHISTICATED INDUSTRIAL BUILDING. NOT SO! IT'S OUR FRIEND THE BEE WHO PRODUCES THIS WONDERFUL SUBSTANCE. THE 12-DAY-OLD BEE HAS WAX-SECRETING GLANDS. A TEMPERATURE OF 97°F/36°C (ON AVERAGE) IS FAVORABLE FOR WAX SECRETION. BUT FOOD ALSO PLAYS AN IMPORTANT ROLE: BEES MUST CONSUME 15 POUNDS (7 KG) OF HONEY TO PRODUCE ABOUT 2 POUNDS (1 KG) OF WAX.

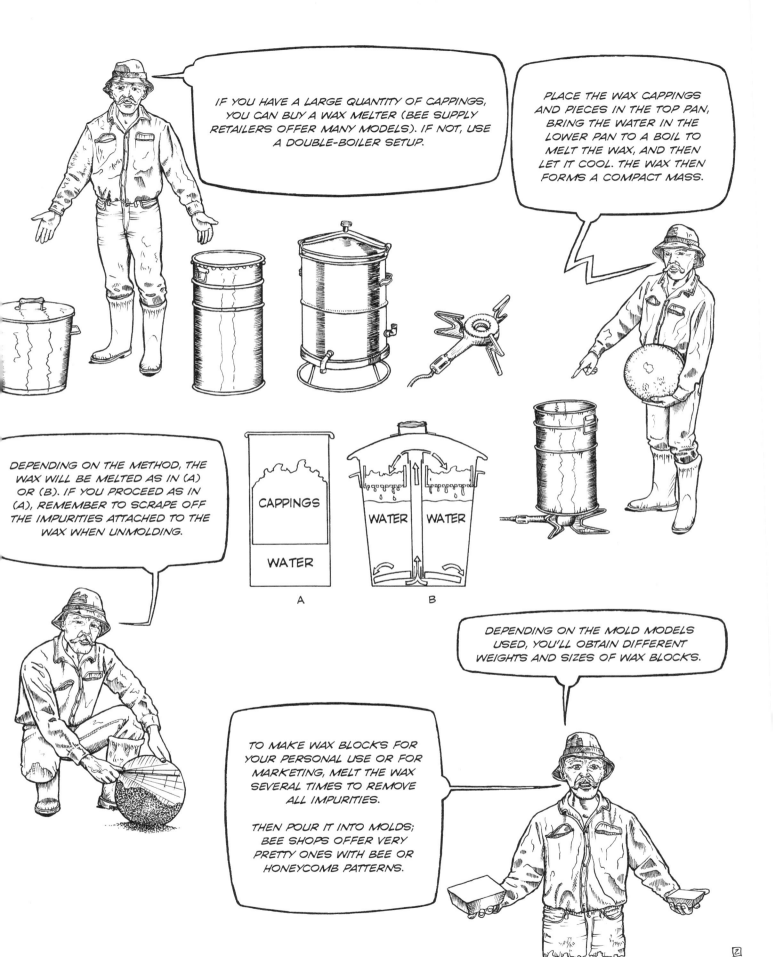

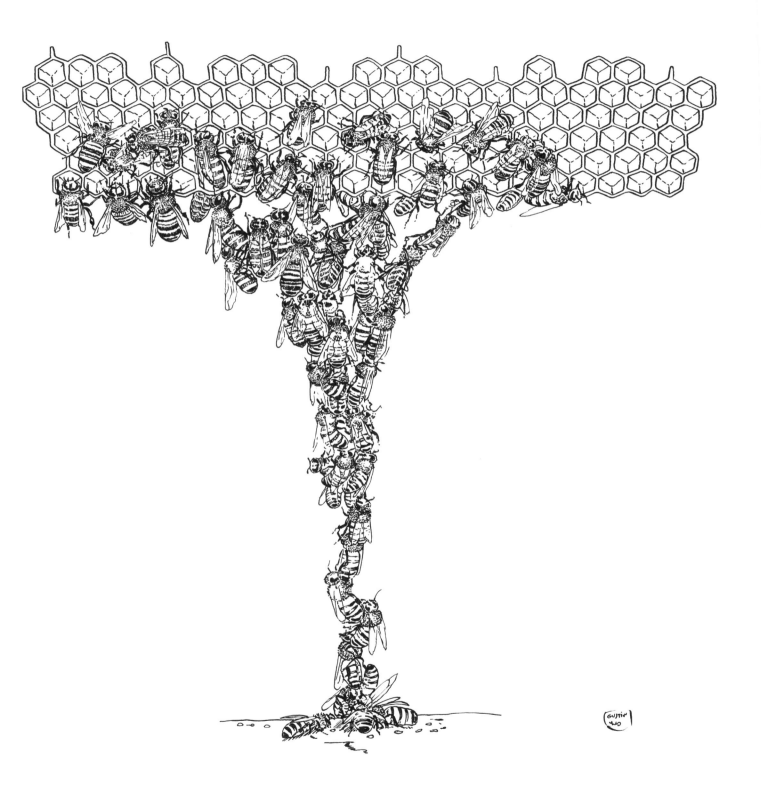

A SOLAR WAX MELTER

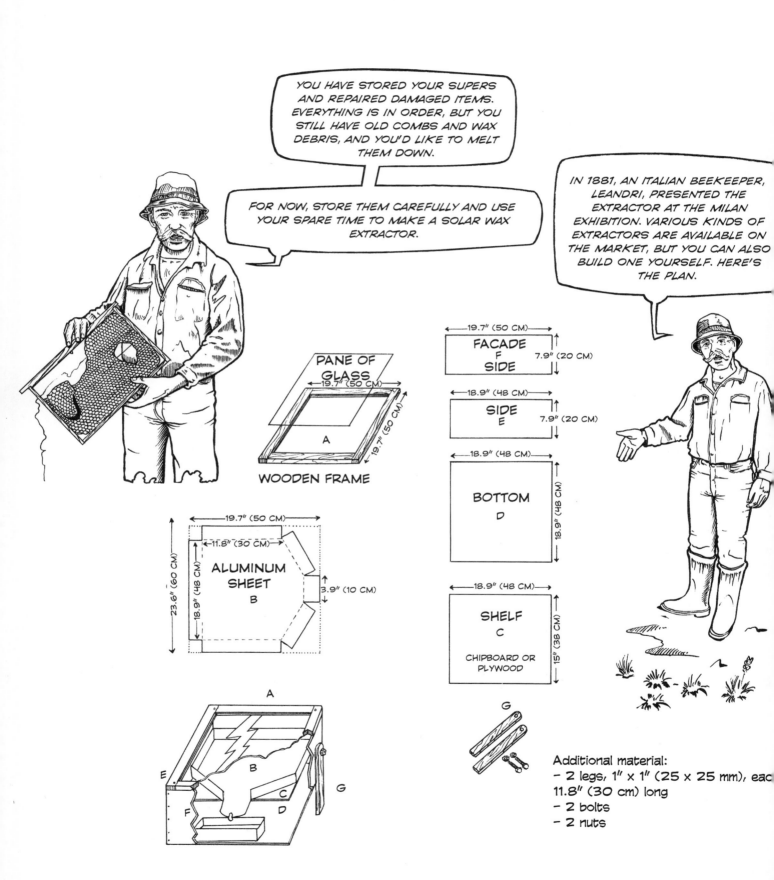

• A SOLAR WAX MELTER •

PROPOLIS

> THE LEAST-TALKED-ABOUT HIVE PRODUCT IS PROPOLIS, ALTHOUGH IT'S VALUABLE. MANY BEEKEEPERS IGNORE THE ESSENTIAL CONTRIBUTION OF PROPOLIS; THEY REJECT IT AND NEGLECT IT. HAVE YOU EVER FOUND A FRAME THAT'S SO DIFFICULT TO LIFT THAT YOU ASK YOURSELF IF IT'S BEEN NAILED DOWN? DESPITE THIS INCONVENIENCE BOTH FOR THE OPERATOR AND THE OCCUPANTS OF THE HIVE, A PROPOLIZED HIVE IS A SIGN THAT IT'S "HEALTHY," AND IT'S ALSO USEFUL WHEN TRANSPORTING A HIVE.

Egyptians have used propolis for embalming the dead since antiquity. The Romans also used it. In the past, it was used to varnish musical instruments and other objects, and also as a grafting putty. Farmers used it when a cow had a broken horn, and it healed without any problems.

Propolis is harvested during the hottest hours of the day. Bees collect sap from the buds of certain plants, bring it back to the hive, modify it by adding saliva secretions, and make it into propolis.

A

B

C

> LET'S WATCH A BEE (A) LOCATE A POPLAR BUD (ALDER, HORNBEAM, ETC.) AND FILL ITS BASKETS IN THE SAME WAY AS IT DOES WHEN IT STORES POLLEN. IT USES ITS MANDIBLES AND FRONT LEGS TO RECOVER THE RESIN (B), WHICH IS REDUCED BY KNEADING, THEN IS PLACED IN THE BASKETS. WHEN THEY'RE FULL, THE BEE RETURNS TO THE HIVE. WITH THE HELP OF A WORKER (C), SHE GETS RID OF HER LOAD. IT'S IMMEDIATELY USED TO SEAL CRACKS OR FOR ANY OTHER MASONRY WORK.

> OK! LET'S GET STARTED.

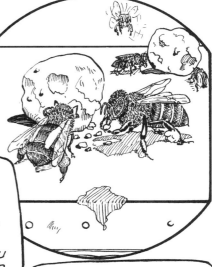

BEES DON'T COLLECT MATERIALS TO MAKE PROPOLIS ONLY FROM BUDS. AS PROOF: ONE DAY WHEN I WAS SCRAPING SOME FRAMES, I WAS ABLE TO MAKE SMALL PROPOLIS BALLS, WHICH I PLACED ON THE ROOF OF AN EMPTY HIVE.

I NOTICED THAT THE BEES CAME AND BROKE UP THESE LITTLE BALLS AND LEFT LOADED WITH PROPOLIS. IN LESS THAN TWO DAYS, THEY HAD TAKEN EVERYTHING! IF YOU CAREFULLY OBSERVE YOUR BEES, YOU'LL NOTICE THAT THEY TAKE PROPOLIS WHERE THEY FIND IT (OLD FRAMES, OLD HIVES, BEE WASTE, ETC.).

IT WAS ONCE SAID THAT BEES FOLLOWED THEIR DEAD MASTER'S COFFIN, NOT OUT OF AFFECTION BUT TO COLLECT THE VARNISH FROM IT.

COMPOSITION

Resinous materials (sap)
Wax
Pollen
Balms
Vitamins
Antibiotic substances

PROPERTIES

Healing
Antibiotics
Antiviral
Antimicrobial properties

DEAR READER, TAKE A LOOK AT THE COMPOSITION AND PROPERTIES OF PROPOLIS. DON'T BE SURPRISED WHEN YOU SEE US SEARCHING FOR THIS WONDERFUL SUBSTANCE!

MAYBE YOU THINK WE'D BETTER GO FORAGE FOR NECTAR RATHER THAN MAKE PROPOLIS. WELL, YOU'RE WRONG, BECAUSE PROPOLIS IS VERY USEFUL IN OUR HOME.

WE USE IT TO SEAL CRACKS OR HOLES IN THE HIVE, WELD FRAMES, AND REDUCE THE ENTRANCE. WHEN WE HAVE UNWANTED GUESTS, SUCH AS MICE, WE INJECT THEM WITH A DOSE OF VENOM TO KILL THEM; THEN WE COAT THEM WITH PROPOLIS SO THAT THEIR BODIES DON'T DECOMPOSE.

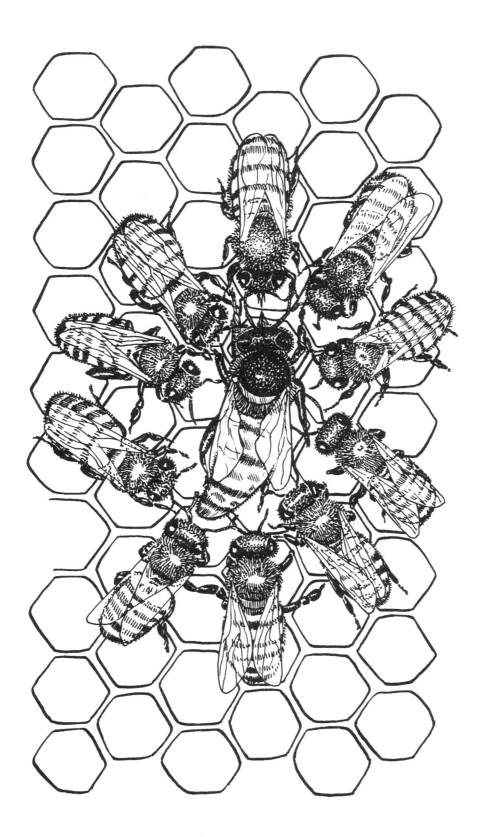

POISONING

I HOPE YOU'LL NEVER BE CONFRONTED WITH THE NEED TO DEAL WITH A POISONING SITUATION. BUT, IF NECESSARY, YOU'LL ALSO WANT TO CONSIDER FILLING OUT AN INSURANCE CLAIM.

NOTHING IS MORE UNPLEASANT OR SADDER THAN TO BE ROBBED OF YOUR BEES. TO SEE YOUR BEE POPULATION REDUCED BY HALF OR EVEN TO NOTHING HAS HUGE EMOTIONAL IMPACT.

IN TERMS OF THE PRACTICAL MEASURES, TAKE OUT THE NECESSARY INSURANCE POLICIES TO PROTECT YOURSELF; SEE WHAT YOUR INSURANCE COMPANY RECOMMENDS: CIVIL LIABILITY, THEFT, FIRE, VANDALISM ...

COME ON, COME ON! LET'S HURRY! SPRING IS HERE ... BUT BEWARE, SO ARE POISONOUS SUBSTANCES!

DISEASES

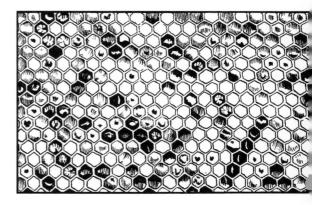

THERE ARE DIFFERENT KINDS OF DISEASES: THOSE THAT ATTACK THE BROOD, THOSE THAT INFECT ADULTS, AND, FINALLY, THOSE THAT ATTACK BOTH. WE WILL TALK HERE ABOUT ONLY SOME OF THE MOST COMMON DISEASES, BUT MANY OTHERS ARE FOUND IN APIARIES.

DO YOUR RESEARCH! WE'LL ONLY BRIEFLY MENTION A FEW DISEASES HERE. USE RELIABLE AND CURRENT SOURCES, SUCH AS YOUR REGION'S BEE INSPECTOR, TO STAY INFORMED ON PROBLEMS AND SOLUTIONS.

PREDATORS, PARASITES, AND NOW DISEASES. WHAT'S NEXT? NUCLEAR WAR?

COME ON, FRED; CALM DOWN!

FOULBROOD

These are microbial diseases that cause a lot of damage in hives. A distinction is made between American and European foulbrood.

Detail of a comb infected by foulbrood. It's irregular; the cells are collapsed or pierced.

AMERICAN FOULBROOD (AFB) IS A HIGHLY CONTAGIOUS DISEASE CAUSED BY A MICROBE, BACILLUS LARVAE, WHICH ATTACKS THE BROOD AT ALL STAGES OF ITS DEVELOPMENT. THE SIGNS OF THIS DISEASE ARE PARTICULARLY IDENTIFIABLE: THE BROOD IS IN AN IRREGULAR PATTERN; THE CAPPED BROOD IS PIERCED OR PARTLY UNCAPPED BY BEES SEEKING TO EXTRACT DISEASED LARVAE. THE SICK BROOD HAS A SMELL SIMILAR TO STRONG GLUE. THE LARVAE BECOME VISCOUS, AND WHEN YOU TRY TO REMOVE THEM, THEY STRETCH INTO SLIMY STRINGS. THE DISEASE IS TRANSMITTED BY THE INGESTION OF HONEY INFESTED WITH BACILLUS SPORES.

I'M SICK ...

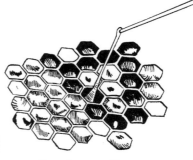

The diseased larva extracted from its cell is slimy and stringy.

EUROPEAN FOULBROOD IS CAUSED BY SEVERAL MICROBES: MELISSOCOCCUS PLUTONIUS, STREPTOCOCCUS APIS, BACILLUS ALVEI, BACILLUS ORUS, AND BACILLUS EURIDICE. THIS DISEASE OCCURS IN LARVAE, WHICH BECOME TRANSPARENT AND THEN, AS THE INFECTION PROGRESSES, TURN YELLOW TO A BLACKISH TINGE. THE LARVAE NO LONGER ADHERE TO THE BOTTOM OF THE CELLS AND GIVE OFF A ROTTED SMELL. UNLIKE AMERICAN FOULBROOD, THE LARVAE AREN'T STRINGY. EUROPEAN FOULBROOD IS LESS SERIOUS THAN AMERICAN FOULBROOD.

I DON'T FEEL GOOD...

CHRONIC BEE PARALYSIS VIRUS (CBPV)

A CONTAGIOUS DISEASE CAUSED BY VARROA MITES. SMALL BLACK BEES SHOWING NERVOUS TREMORS THAT CAUSE PARALYSIS AND DEATH CHARACTERIZE THIS DISEASE.

THE EASIEST WAY TO FIGHT CBPV IS TO CHANGE THE HIVES' LOCATION SO THAT THE BEES FIND A BETTER SOURCE OF NECTAR.

NOSEMA

This contagious disease is due to a fungus, *Nosema apis*, which is present in the bee's digestive tract. Contagion is caused by spores present in the droppings of sick bees. The disease spreads through honey, pollen, and wax. An entire apiary can be affected by nosema; robbers who steal the honey from the sick and weakened hive will spread it.

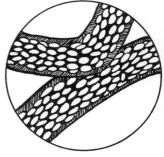

Nosema spores in a bee's organs

THIS DISEASE IS MOST OFTEN FOUND IN THE SPRING, AND IT'S A MAJOR CAUSE OF MORTALITY AMONG BEES.

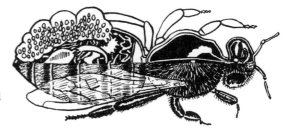

Bee's intestine infested by nosema

STONEBROOD

Stonebrood is usually caused by *Aspergillus flavus*, a fungus that affects both brood and adult bees. On the frames, mummified brood patches can be seen. In adults, the fungus first attacks the intestine and then spreads throughout the body. Unable to fly, the bees die in front of the hive.

CHALKBROOD

Chalkbrood is caused by the fungus *Ascosphera apis*, which attacks the brood and causes death. The larvae take on a chalky appearance; hence the name. Good ventilation of the hives and a sunny location will limit chalkbrood conditions.

Mummified brood

LET ME SAY IT ONE MORE TIME: I'VE PURPOSELY NOT MENTIONED TREATMENTS FOR THESE DISEASES, SINCE LAB RESEARCH IS PROGRESSING FROM YEAR TO YEAR. IF ONE OF YOUR HIVES HAS A DISEASE, CONTACT YOUR AREA'S BEE INSPECTOR FOR HELP. THEY CAN ADVISE YOU ON TREATMENT.

PARASITES

THIS UNATTRACTIVE SUBJECT IS KEY FOR YOU TO LEARN ABOUT, BECAUSE THE INSIDE OF THE HIVE IS A FAVORABLE ENVIRONMENT (TEMPERATURE, HUMIDITY, ENCLOSED SPACE) FOR THE DEVELOPMENT OF PARASITES. THEY CAN ATTACK BROOD AND ADULT BEES AND SOMETIMES EVEN DESTROY AN ENTIRE COLONY.

THE BEST-KNOWN PARASITES
(there are others)

BRAULIDAE (BEE LOUSE)   ACARAPIS WOODI (TRACHEA MITE)   VARROA MITES

THE BEE LOUSE

Braulidae can be identified by its red color. Three pairs of legs attached under the abdomen and equipped with small feathers allow it to move with great agility on the moving bee.

This dipteral can be seen by the naked eye, since it's 1 mm wide. It's usually found hanging from the worker's corselet, but its favorite host is the queen, who can support about 30 individuals. When the number of lice is not too high, bees don't care because they don't directly harm their health. To feed, *Braulidae* lice simply suck the drops of honey that flow during a feeding among workers, bees, and queen, then return to their place.

HOW TO CONTROL THIS PEST

Here's a very simple method: put a handful of tobacco in your smoker, put a cardboard box (or varroa tray) on the hive tray, and smoke the colony. Wait a moment, remove the cardboard on which the lice have fallen, and burn it away from the hives. To be extra careful, you can repeat the operation a few days later. Beekeepers who treat for mites preventively, of course, will be less likely to encounter this parasite.

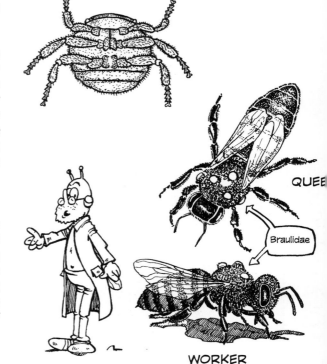

QUEEN — Braulidae

WORKER

ACARAPIS WOODI

Acariasis or tracheal mite was originally called the Isle of Wight disease because it was there that the disease first appeared, according to English reviews dating back to 1904. Fifteen years later, researchers discovered the cause of this parasitic disease in a laboratory in Scotland. It is a mite invisible to the naked eye, located in the tracheas of the bee, and causes death by asphyxiation.

LARVA — MALE — FEMALE

Acariasis infects young bees aged nine to ten days. Mites evolve in the first two tracheas of the bee and pierce them to suck blood.

Mating and egg laying take place inside the tracheas. The female lays about half a dozen eggs that turn into larvae and then perfect insects that can then reproduce; the cycle will have lasted only 15 to 20 days. Quickly, the droppings and waste of the growing number of *Acarapis woodi* block the tracheas. The bee dies.

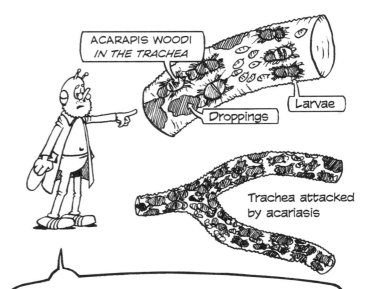

ACARAPIS WOODI IN THE TRACHEA — Droppings — Larvae

Trachea attacked by acariasis

THE ACARAPIS WOODI MALE HAS A SMALLER ABDOMEN THAN THE FEMALE. IT'S EQUIPPED WITH FOUR PAIRS OF LEGS. THE FIRST PAIR HAS HOOKS; THE SECOND AND THIRD, SUCTION CUPS; AND THE FOURTH, LONG HAIR.

IN SPRING, BEES UNABLE TO FLY AND FALLING IN FRONT OF THE HIVES CHARACTERIZE AN APIARY AFFECTED BY ACARIASIS.

ACARAPIS WOODI TOOK ADVANTAGE OF THE WINTER TO DEVELOP AND MANIFEST ITSELF IN THE SPRING. TO FIGHT THIS MITE, YOU'LL FIND MANY EFFECTIVE SMOKE PRODUCTS ON THE MARKET. BUT BEFORE TRYING ONE OUT, DON'T HESITATE TO ASK YOUR BEE INSPECTOR FOR ADVICE.

VARROA MITES

Varroa is made up of two species: *Varroa jacobsoni* and *Varroa destructor*. This mite was discovered on the island of Java at the beginning of the 20th century by Professor Edward Jacobson. The plague quickly spread to Europe, Africa, and South America. It's difficult to recommend processes to control varroa mites because they are constantly evolving. Bee supply retailers have equipment and products to treat for varroa.

Varroa destructor, visible to the naked eye, has four pairs of legs. The female is reddish brown and measures 2 mm. The male, yellowish white in color, is much smaller. Varroa mites attack both larvae and adult bees and feed on their blood. The female varroa lays its eggs on the bee larvae just before capping. Fertilization takes place inside the cell. Male eggs hatch after six to seven days, female eggs after about eight days. When the young bee leaves its cell, it will already have one or more fertilized females attached to it. It should be noted that the varroa particularly likes drone brood.

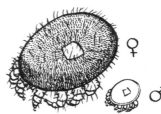

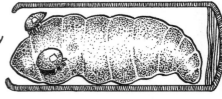

Bees affected by varroa mites

Female varroa mites lay their eggs before capping.

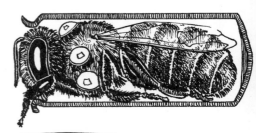

An infested drone leaving its cell

> THE VARROA MITE LIKES THE AMBIENT TEMPERATURE OF OUR HIVE, BUT IF THE TEMPERATURE IS ABOVE 104°F (40°C), IT DIES.

> AMAZING! JUST DRYING MY HAIR KILLED IT.

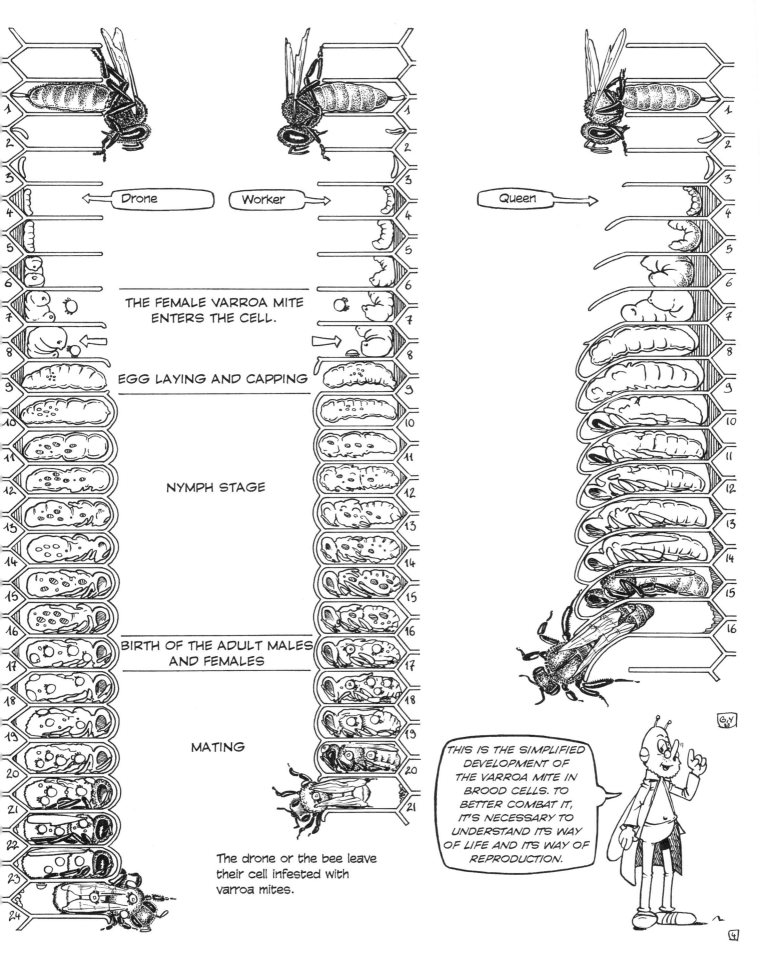

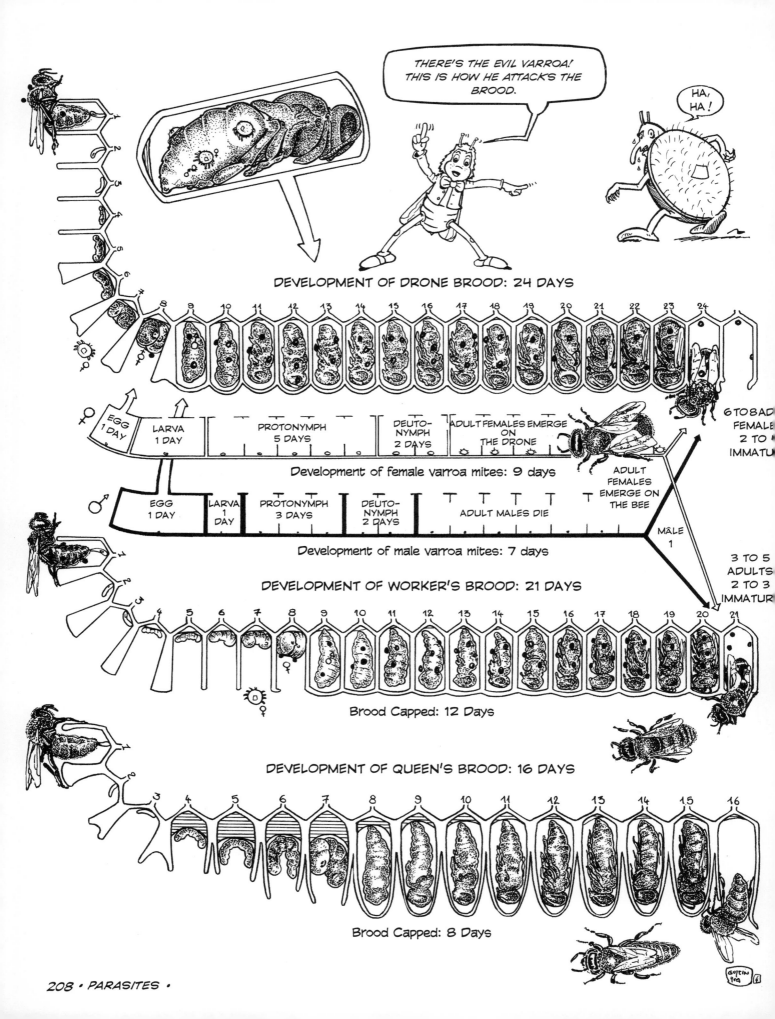

WAX MOTHS

IF YOUR HIVE IS UNINHABITED FOLLOWING THE DEATH OF A COLONY, DON'T LEAVE IT AS IT IS—IT WOULD VERY QUICKLY BE INVADED BY THE GREATER WAX MOTH.

THERE ARE DIFFERENT VARIETIES. THOSE THAT INTEREST US ARE GALLERIA MELLONELLA AND MELIPHORA GRISELLA. THE WAX MOTH ENTERS THE HIVE MAINLY AT NIGHT AND PREFERS A WEAK COLONY. IT USUALLY LAYS EGGS ON COMBS CONTAINING POLLEN BECAUSE, TO PRODUCE SILK, THE LARVA NEEDS TO FEED ON NITROGEN-RICH MATERIALS.

LARVA

A SINGLE MOTH ENTERING A HIVE CAN DESTROY AN ENTIRE COLONY BECAUSE OF ITS RAPID REPRODUCTION. IT CAN LAY 200 EGGS. IN THE SECOND GENERATION, THERE WILL BE SEVERAL THOUSAND INDIVIDUALS. IN THE THIRD GENERATION, OVER A MILLION. ALSO, THE FEMALE CAN LAY EGGS WITHOUT HAVING BEEN FERTILIZED. WEAK COLONIES HAVE DIFFICULTY GETTING RID OF SUCH AN ENEMY BECAUSE, TO MOVE ON THE COMB, THE WAX MOTH BUILDS SILK TUNNELS, WHICH BEES FEAR GETTING THEIR LEGS TANGLED UP IN.

OPEN AND CLOSED CHRYSALIDES

WHEN EXAMINING A FRAME, IF YOU NOTICE THE PRESENCE OF TRACKS, THEY ARE TUNNELS OF THE GREATER WAX MOTH.

EGGS

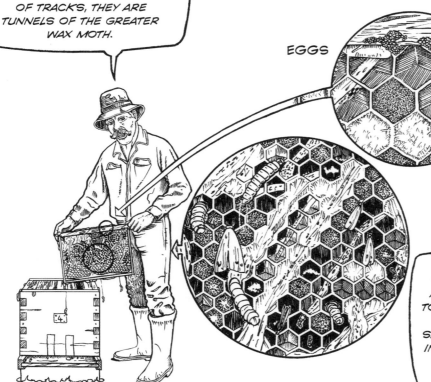

SUPER FRAMES AREN'T IMMUNE TO THIS PREDATOR. PRECAUTIONS SHOULD BE TAKEN IN EARLY AUTUMN.

GLOSSARY

A

ABDOMEN: The posterior part of the bee in which the crop, stomach, small intestine in the worker, and genital organs in the queen and males are located.
ACARIASIS: Disease caused by a microscopic mite of the species *Acarapis woodi*
AMBROSE: Patron saint of beekeepers
ANTENNAE: Organs that communicate, feel, and measure temperature and humidity as well as vibrations. The bee's radar.
APIARY: Set of hives; the location of the hive is essential for the proper functioning of the colonies.
AWNING: Eaves allowing the near impermeability of the bottom board. It can be fixed or removable.

B

BEARDING: In spring, a large number of bees hang in front of the hive due to lack of air or space.
BEE BRUSH: A beekeeper needs a bee brush at harvest time to rid the frames of the remaining bees. It must not be too hard or too flexible to be effective.
BEE DANCE: No, no, don't get me wrong—they're not going to a dance! For them, it's a kind of language: they dance to point their colleagues in the direction of a nectar source.
BEE ESCAPE: Device placed between the super and the body of the hive used to let the bees out and prevent them from accessing the super, to facilitate the beekeeper's harvesting work
BEE GLANDS: The three main ones are salivary glands, wax glands, and venom glands. Each of them has a specific function, as their name suggests.
BEE LOUSE: *Braulidae*; parasitizes the bee by clinging to it
BEE-EATER: Insect that feeds on bees
BREEDING: Breeding queens allows the beekeeper to make a good selection, replace elderly queens, or requeen a queenless colony. This operation requires a lot of knowledge about bees.
BROOD: All the eggs and larvae found in the cells that have been laid by the queen

C

CANDY: A solid sugar and honey preparation, which is an excellent stimulant in spring for bees but also in autumn to supplement winter supplies
CAP: Wax film that covers the honey cells. This cover is waterproof, white in color. The cap covering the brood is permeable and brown in color; this difference is explained by the fact that honey must be hermetically sealed while the brood must breathe.
CELL: On one comb are those for workers and those for males. The latter have larger dimensions, and in some periods (swarming), there may be one or more acorn-shaped queen cells.
CLIPPING: Cutting off one or both of the queen's wings to prevent the swarm from flying away from the hive after her
COLONY: Group of bees composed of 40,000 to 60,000 workers, 1,000 to 2,000 males (or drones), and a queen.
COMB: All the cells together
CROP: Nutrition and processing organ, also called a honey stomach. Its function is remarkable: the bee collects nectar and stores it in the crop, then, by a system of pharynx-crop comings and goings, mixed with diastase, it's transformed into honey, then disgorged into the cells. However, it can be assimilated by the bee in order to support itself.

D

DADANT, Charles (1817—1902): He made modifications to hive structure that gave a very satisfactory result.
DRIFTING: Even with good orientation, bees arriving at the apiary can enter other hives, especially when they are poorly placed.
DRONES: Males, much larger than the worker; their essential role is to fertilize the queen. At the end of the summer, they are expelled from the hive by the workers.

E

EGG: It's small (length 1 to 1.5 mm). During egg laying, it's placed perpendicular to the bottom of the cell, and day after day its position changes. On the fourth day, it's called a larva.
ENTRANCE: Careful observation of the entrance lets a beekeeper know what is happening inside the hive, if the colony is in good health, and whether or not to open the hive to examine the frames in more detail. It is located at the bottom of the hive. It allows the entry and exit of bees.
ENTRANCE REDUCER: As its name suggests, it reduces the size of the entrance to the hive and is placed in the fall to avoid the intrusion of some predators during the winter.
EQUIPMENT: Essential for the beekeeper who visits the apiary: smoker and fuel, a hive tool, a veil (for facial protection)
EXTRACTOR: Manual or motorized device used by a beekeeper to extract honey from the frames. There are different types of extractors: tangential and radial.
EYES: Bees' eyes are composed of three simple eyes called ocelli and two compound faceted eyes.

F

FAN BEES: Worker bee that uses its wings at the entrance and inside the hive, to cause a slight draft that helps cool the temperature, reduces the humidity level, and allows good ventilation.
FEEDER: Its function is obviously to provide food (complementary or stimulating) to bees. There are many models on the market, and there are many creative solutions for DIY-ers.
FEEDING: There are two types: the one to supplement insufficient provisions for a good wintering, and the one to stimulate the queen's egg laying in early spring.
FOLLOWER BOARD: Mobile partition of frame size, made with a wooden board and used to isolate part of the hive
FORAGER: In the life of a bee, there are several stages, and, among them, the one that occurs around the 21st day is collecting nectar; hence its name.
FOULBROOD: Contagious disease caused by a microbe that breaks down the brood. There are two types: European and American foulbrood.
FRAME: Assembly of slats of different sizes according to the model of the hive used, and reinforced with wire on which foundation wax is embedded. They are placed in the body of the hive, respecting the space necessary for the passage of bees. These frames allow bees to build combs that can easily be manipulated.
FRISCH, Karl von (1886—1982): Thanks to his careful observations, he provided important clarifications concerning the meaning of the bees' dances.

G

GRAFTING TOOL: Small tool used for queen breeding. One of its ends, in the form of a spoon, is used to collect the royal jelly, and the other, in the form of a hook, allows the larvae to be gently collected and placed in the bottom of the cup.

GRANULATION: Refers to the solid consistency that honey takes after extraction. Not all honeys granulate, such as honey from the flowers of linden or acacia.

GREATER (OR LESSER) WAX MOTH: *Lepidoptera*. This small moth, when it enters a hive, causes considerable damage. Supers stored by the beekeeper are not safe from this enemy unless necessary precautions are taken.

GUARD BEE: After having been a wax maker, the 16-to-20-day-old bee watches and guards the entrance to the hive.

H

HIVE: Houses a colony of bees. There are two types of hives: fixed hives (baskets, tree trunks, etc.) and hives with movable frames (Dadant, Langstroth, etc.).

HIVE REGISTRATION: In the US, honeybee hives, both commercial and hobbyist level, are monitored by the Department of Agriculture, and states vary in their requirements for registering hives. Check your state's and local government's requirements.

HIVE STAND: An element that we tend to neglect, yet of great importance, because it protects the hive (e.g., from humidity, predators). Various stands exist: wooden bases, tubular, tires, blocks, etc.

HIVE TOOL: An essential tool for a beekeeper. There are different models, with claws and springs, but a basic one, or a wood chisel, also does the job very well. It's used to remove a super or a frame (or both) during a visit.

HONEYDEW: Sweet liquid expelled by aphids or exuded by certain plants and collected by bees

HUMIDITY: The concentration of bees inside the hive leads to high humidity; the beekeeper will try to remedy this by placing a blanket (burlap) on the frames to absorb excess water vapor; too-high humidity would cause considerable damage to the bees.

I

INNER COVER: The hive's ceiling; it can be made of wood, plastic, or canvas. It protects the frames.

INTOXICATION: This word can be applied to bees when they benefit from an abundant nectar source (honeydew on canola, for example).

INTRODUCING A QUEEN: An old queen or a queenless colony requires you to introduce a queen. Several methods are possible, the most commonly used being perhaps the use of a queen cage, but experience will help you decide which method is most effective for you.

ITALIAN: A breed of bee that's very pleasant to work with. The crossing of the Italian and the black bee has produced a race with a defensive tendency.

J

JANUARY: This month's work consists of visits to the hives to observe the entrances and to make sure that nothing abnormal is happening there. Apart from these visits, you can repair and repaint hives and assemble frames.

JOINING: Also called combining. This operation consists of bringing together two weak colonies or a queenless colony with a small population colony.

JULY: Whatever region you live in, July will require your full attention; in a honey-producing region you'll be laying and removing honey supers; in a region where the flora is not sufficient, you need to provide food for your bees.

JUNE: You'll be very busy this month. First, if you have placed supers for a good honey flow, you'll have to remove them (as soon as the frames are capped). Second, this month is a good time for making splits.

L

LANGSTROTH, Lorenzo (1810—1895): He was a teacher, then a pastor. His passion for beekeeping began in his retirement. The new elements he brought revolutionized beekeeping: everyone knows of the Langstroth hive.

LANGUAGE: This term attributed to bees may come as a surprise; yet, the proper functioning of a colony is due to communication, done either by special flight patterns or by a noise imperceptible to our ear, but which is their own language.

LARVA: Name given to a four-day-old egg. The larvae are fed the first three days with royal jelly, then mainly with a food including nectar and pollen.

LOCATION OF THE HIVES: The most favorable spot is one that is not humid, is protected from high winds, gets morning sun, and has a natural or artificial source of water.

M

MEAD: This honey-based drink is delicious and was first enjoyed centuries ago.

N

NASONOV GLAND: Located on the upper part of the abdomen, it's a pheromone-emitting organ for communication.

NECTARY GLAND: Gland located on certain plants that secretes a sweet liquid that bees appreciate (nectar)

NOSEMA: A disease that affects the stomach of adult bees and is caused by a parasite called *Nosema apis*

NUC: A mini-beehive containing five frames, which is temporarily used to house a swarm. There are also breeding nucs and mating nucs (nuclei).

NURSE BEE: Bee aged from five to six days. Its job is to feed the larvae.

P

PHEROMONES: Volatile substances allowing communication between insects. It's the bees' ID card.

POLLEN: Substance comparable to grains of dust, which is the male element of a flower and is harvested by the bee. Pollen is essential for workers and necessary for feeding their larvae. Insects and especially bees promote pollination by transporting pollen to the female organs of the same flower or another.

POLLEN BRUSH: Allows pollen to be collected, thanks to the hairs on the bee's legs

POLLINATION: Transport of pollen from the androecia to the gynoecia. Pollination is direct when it occurs on the same flower and indirect when it occurs from one flower to another.

PROPOLIS: A kind of resin made by bees from the sap of certain trees. Bees use propolis to seal cracks and holes in the hive. The therapeutic properties of this resin are very interesting.

Q

QUEEN: She has an essential role, that of laying eggs. When she's no longer able to perform this function, she is replaced naturally or artificially.

QUEEN BEE BALLING: Occurs when the queen is surrounded by enraged bees trying to kill her, especially after the introduction of a queen
QUEEN CAGE: Mesh box allowing queens to be transported without risk. Also used for the introduction of queens.
QUEEN CUP: Artificial cells made of plastic or wax.
QUEEN EXCLUDER: This is a grid, placed between the hive body and the supers, that lets worker bees into the supers but not the queen. This is to stop the queen from laying eggs in the supers.
QUEEN MARKING: You alone will decide whether or not to mark your queens (some are reluctant to mark them), either with a paint spot or an adhesive pad (colors change depending on the year). Marking makes it easier to identify the queen.
QUEENLESS COLONY: A colony that has lost the queen. The beekeeper notices this during the first spring visits. The causes are diverse. She may have been crushed during a visit, she may have died during the winter, she may have died of old age, etc.
QUIMBY, Moses (1810–1875): An American beekeeper; he was the inventor of the hive that bears his name, a model similar to the Langstroth and whose frame dimensions are 46 x 27 cm. The idea of the partition in the hive is also his. And let's not forget the smoker; it was Quimby who invented it.

R

REQUEENING: Introducing a new queen
ROBBING: Attack mounted against a weak colony by bees that come in force to seize the food from the hive.
ROOF: Depending on the beekeeping method chosen, it will be a chalet roof or a flat roof; several materials can be used for their manufacture (wood or straw, wood covered with aluminum, plastic, zinc, etc.).
ROYAL JELLY: It contains many chemical substances and fats; it's the food of the queen larvae. This is not its only use, because some beekeepers benefit from its virtues by eating a tiny amount.

S

SUCROSE: Chemical name for sugar. The nectar harvested by bees contains a tiny amount of it. The sucrose content of nectar varies depending on the plant.
SECTION: Small plastic or wooden box, without bottom or lid, that is placed in the hive to collect small honeycombs. This presentation of honey on the comb is very attractive and a much-appreciated gift.
SELECTION: This allows the beekeeper to have almost perfect colonies (i.e., hives with fertile queens and gentle, hardworking bees).
SENSES: The bee has several: orientation, smell, touch, sight, and hearing. It should be added that the bee has an excellent memory and a very good sense of time.
SEPTEMBER: This month, depending on the region, will be devoted to harvesting or getting your hives ready for winter.
SMELL: The most developed of the senses that bees possess. You can notice this during a visit: the unpleasant smell of excessive perspiration or heavy perfume irritates them and makes them aggressive.
SMOKE: Before opening a hive, a little smoke is puffed inside to calm the bees. The fuel used may be burlap cloth, wood chips, or dry, dead leaves. The important thing is that the smoke is cold (white).
SMOKER: It could be nicknamed "the essential" because when used wisely, it's an essential accessory for the beekeeper. The smoke emitted by the smoker must be white and cold, so be careful about the fuel used.
SOLAR WAX MELTER: Box fitted with a pane of glass, used for melting wax slowly in the sun
STARTER STRIP: A strip of foundation wax that is inserted into the groove of the frame and sealed in with a drop of wax

STINGER: It's the bee's defense. To get rid of an opponent, a bee injects it with venom. Males have no stinger.
STRAINER: A honey strainer is essential when you extract your honey because it removes impurities from the honey such as wax debris, caps, etc.
SUGAR: While it's preferable to consume honey rather than sugar, let us not denigrate the latter because it's necessary for the manufacture of syrup and candy.
SUPER: It has the same dimensions as the body of the hive in terms of length and width. The height is in principle reduced by half (but this depends on the type of hive used). The installation of a super is justified when a honey flow is imminent, when the population is large and occupies almost the entire body.

T

TAPPING (also called drumming): The act of hitting the walls of the hive or the basket to move the bees up either into a box or into a hive to harvest a swarm
TEMPERATURE: Temperature variations play a major role in the activities and work of bees. The temperature favorable to the proper functioning of a colony can be considered to be about 77°F (25°C) outside and 97°F (36°C) inside the hive. The temperature also influences the queen's egg laying and the hatching of the brood.
TONGUE: Without this organ, how would bees harvest nectar? The length of their tongue allows them to forage a greater or lesser number of flowers.
TOOLS: Among the essentials are a smoker, a hive tool, and a bee brush.
TRANSFERRING: Process used by the beekeeper when he or she wants their bees to have a comfortable home. The total content (honey, brood, and bee frames) of a fixed hive (tree trunk, basket, etc.) is transferred to a frame hive. There are various methods of transferring.
TREATMENT: Before any treatment, don't hesitate to ask a bee inspector to look over your hives; he or she will be able to advise you.

U

ULTRASOUND: Emitted by the flapping of the bees' wings, it's part of their communication system.
ULTRAVIOLET: Complementary color to the blue green perceived by bees.
UNCAPPING: Action carried out by a beekeeper to remove the thin layer of wax deposited by the bees on the honey cells
UNCAPPING KNIFE: Knife used to uncap the wax cells at harvest time
USEFULNESS: Bees are useful to everyone. The farmer who owns a field will see their yield considerably improved. The beekeeper will be satisfied with a good honey harvest, and the consumer will enjoy good, healthful honey.

V

VARROOSIS: Parasitic disease caused by the **varroa** mite that affects bees at all stages of their development. This parasite is a real scourge for the beekeeper.
VEIL: Very useful for protecting your face against bee stings
VENOM: Toxic substance secreted by two glands and injected by the bee's stinger
VENTILATION: Necessary for bees in summer in hot weather, but also in winter. Remember not to cover your hives excessively during the winter period, but beware of predators (e.g., mice).

W

WATER SUPPLY: Even though this term is not specific to bees, it's nevertheless essential. A large container filled with water, on which some twigs or other floating debris (to avoid drowning) is placed in each apiary to provide the necessary water for the bees. With a little imagination, there are different ways to provide them with water.

WAX: Substance secreted by 18-day-old worker bees. The bee uses its hind legs to collect the small scales that form under its abdomen, and chews them before building the cells with its colleagues.

WAX COMB: Found in the hive or attached to a tree trunk by a wild swarm

WAX MAKER: After being nurse bees, 14-day-old bees become wax makers.

WINGS: A bee can fly thanks to the two pairs of wings attached to its thorax.

WINTERING: Bad wintering would be catastrophic; so that winter is not harmful to a colony, you need to ensure that provisions are sufficient (which is sometimes difficult to evaluate, because a mild winter causes excessive consumption). The quality of the wood used to make the hive plays an important role in good or bad wintering.

WORKER BEES: In a colony, they form the majority of the population. During periods of intense activity, their life span is about five to six weeks.

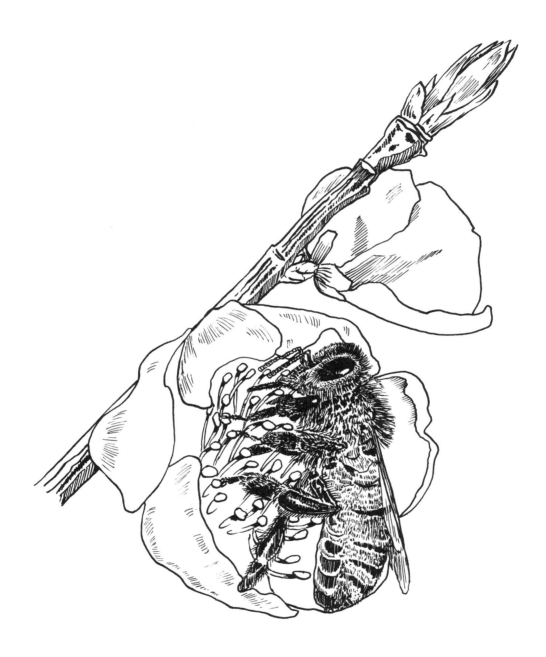

Yves Gustin has enjoyed more than two decades of beekeeping and has learned from bees by watching them carefully. Here he shares his expertise visually with beginners and experienced beekeepers interested in learning some new ways of looking at this fascinating hobby.